Dealing with the Complex Interrelation of Intermittent Supply and Water Losses

Scientific and Technical Report Series No. 25

Dealing with the Complex Interrelation of Intermittent Supply and Water Losses

Bambos Charalambous and Chrysi Laspidou

IWA PUBLISHING

Published by

IWA Publishing
Alliance House
12 Caxton Street
London SW1H 0QS, UK
Telephone: +44 (0)20 7654 5500
Fax: +44 (0)20 7654 5555
Email: publications@iwap.co.uk
Web: www.iwapublishing.com

First published 2017

British Library Cataloguing in Publication Data
A CIP catalogue record for this book is available from the British Library

ISBN: 9781780407067 (Paperback)
ISBN: 9781780407074 (eBook)

Contents

List of Figures

List of Tables

List of Abbreviations

ADD	Average Daily Demand
ALC	Active Leakage Control
AOI	Areas of Interest
AVRL	Aqua Vitens Rand Limited
AWWA	American Water Works Association
CAPEX	Capital Expenditure
DMAs	District Metered Areas
ELL	Economic Level of Leakage
FMO	Dutch Development Bank
FOPIP	Financial and Operational Performance Improvement Plan
GIS	Geographic Information System
HI	High-Income
ILI	Leakage Index
IPCC	Intergovernmental Panel on Climate Change
IWA	International Water Association
IWS	Intermittent Water Supply
JMP	Joint Monitoring Programme
LAMI	Low And Middle Income
MiDA	Millennium Development Authority
MLD	Million Litres per Day
MNF	Minimum Night Flow
MRSP	Metropolitan Region of São Paulo
NGO	Non-Governmental Organization
NRW	Non-Revenue Water
O&M	Operations and Maintenance
OPEX	Operational Expenditure
ORET	Development Related Export Transaction Programme
PIs	Performance Indicators
PL	Physical Losses

PRP	Pressure Reduction Program
PRVs	Pressure Regulating Valves
SC	Service Connections
SIV	System input volume
SNV	Netherlands Development Organisation
SYIP	Subsequent Year Investment Project
THMs	Trihalomethanes
UARL	Unavoidable Annual Real Losses
VUS	Valle Umbra Servizi SpA
WB	World Bank
WLSG	Water Loss Specialist Group
WSI	Water Stress Index

Principal Authors

Bambos Charalambous:
bcharalambous@cytanet.com.cy
Director
Hydrocontrol Ltd
Lemesos
Cyprus

Chrysi Laspidou:
laspidou@uth.gr
Associate Professor
Department of Civil Engineering
University of Thessaly
Volos
Greece

Contributing Authors

Alexandra Spyropoulou:
aespyropoulou@gmail.com
Research Associate
Civil Engineering Department
University of Thessaly
Volos
Greece

Jan Janssens:
jangjanssens2009@gmail.com
Director
JJC sarl
Vercorin
Switzerland

Roland Liemberger:
roland@liemberger.cc
NRW Advisor
Vienna
Austria

Stuart Hamilton:
shamilton@hydrotec.ltd.uk
Hydrotec Ltd
Thorpe Underwood
Northants
UK

Ronnie S Mckenzie:
ronniem@wrp.co.za
Managing Director
WRP Pty Ltd
Pretoria
South Africa

Dimitris Foufeas:
D.Foufeas@olympios.gr
Managing Director
Olympios SA
Athens
Greece

Nicholas Petroulias
np@melcer.gr
Managing Director
Melcer Consulting SA
Athens
Greece

Patricia Bakir:
bakirp@gmail.com
Strategic Planning and Communications Adviser
Water, Energy and Environment Amman
Jordan

Shrihari Sridhar:
hari.sridhar391@gmail.com
Water Loss Expert
Ontario
Canada

Francisco J. Paracampos:
paracampos@sabesp.com.br
Head Central Bussiness Unit
Sabesp
São Paulo
Brazil

Debora Souares:
dsoares@sabesp.com.br
Division Manager
Water Loss Control Division
Sabesp
São Paulo
Brazil

Natally Annunciato Siqueira:
nannunciato@sabesp.com.br
Engineer
Water Supply Operation Division
Sabesp
São Paulo
Brazil

Joe Dalton:
joe.e.dalton@gmail.com
NRW Consultant
Manama
Bahrain

Alessandro Bettin:
alessandro.bettin@hitachi-da.it
Business Development Advisor
Hitachi Drives & Automation (Italy) S.r.l. – Aulos Division
Milan
Italy

Piergiorgio Malano:
joe.malano@hitachi-da.it
Manager Water Network
Hitachi Drives & Automation (Italy) S.r.l. – Aulos Division
Milan
Italy

Acknowledgements

This book is the culmination of efforts and collaborative work of a large group of people that lasted for about two years. Experts in the field were contacted by the principal authors and contributed their valuable expertise and knowledge, making this book a mosaic of inputs that imprints the experience and opinions of a geographically diverse and multi-disciplinary group. We acknowledge the contributions, efforts and support of all the contributing authors and owe them a big thank you for making this book an integrated document that addresses the highly complex issues surrounding Intermittent Water Supply worldwide and its interrelation with water losses.

We would especially like to acknowledge Dr. Alexandra Spyropoulou for her dedication and tireless work on this book; we are grateful for her meticulous calculations on the Global Statistics of IWS, which involved adding populations, country-by-country, in order to reach an estimate of how many people are affected worldwide. Last but not least, we would like to acknowledge Roland Liemberger who kindly accepted to review the manuscript and worked quickly and efficiently so that we can meet our deadlines. His "Austrian-style" somewhat strict but always-constructive criticism throughout the manuscript was extremely valuable and to-the-point and helped us improve the book to its present form.

The Principal Authors

Bambos Charalambous
Chrysi Laspidou

FURTHER ACKNOWLEDGEMENTS

Bambos Charalambous, owner and director of Hydrocontrol Ltd, www.hydrocontrolltd.com, sponsored the colour printing of this book.

Preface

Intermittent Water Supply (IWS) may be defined as piped water supply service that delivers water to users for less than 24 hours in one day.

Although basic issues related to water scarcity and intermittent supply are not part of general public awareness in High Income countries, their effect is truly global as this issue affects over a billion people worldwide adding to the extremely large number of people who do not have any access at all to piped water on premises. It means that around three billion people in the world suffer every day from unsafe or intermittent water supply in their daily lives!

The awareness of world politicians and citizens around this daily shameful reality has developed slowly commencing with the United Nations Millennium Campaign in 2002 which stated: "Eradicating extreme poverty continues to be one of the main challenges of our time, and is a major concern of the international community. The Millennium Development Goals are ambitious but feasible and, together with the comprehensive United Nations development agenda, set the course for the world's efforts to alleviate extreme poverty by 2015".

On 28 July 2010, through Resolution 64/292, the United Nations General Assembly explicitly recognized the human right to water and sanitation and acknowledged that clean drinking water and sanitation are essential to the realization of all human rights.

The Resolution calls upon States and international organizations to provide financial resources, help capacity-building and technology transfer to help countries, in particular developing countries, to provide safe, clean, accessible and affordable drinking water and sanitation for all.

The water supply for each person must be sufficient and continuous for personal and domestic uses. These uses ordinarily include drinking, personal sanitation, washing of clothes, food preparation, personal and household hygiene. According to the World Health Organization (WHO), between 50 and 100 liters of water per person per day are needed to ensure that most basic needs are met and few health concerns arise. Everyone has the right to a water and sanitation service that is physically accessible within, or in the immediate vicinity of the household, educational institution, workplace or health institution.

Bambos Charalambous and Chrysi Laspidou
doi: 10.2166/9781780407074_FP025

According to WHO, the water source has to be within 1,000 meters of the home and the collection time should not exceed 30 minutes. Can we be happy with that? Surely not and having the possibility to get water at the tap is clearly an objective!

In September 2015 the UN Sustainable Development Goals have put water and waste water in the light again, with in particular specifics goals "6" related to water and waste water. The SDG's pragmatically still do not specify that water should be at the tap and permanent, but we need to stress that permanent supply is a key success factor for most of the other goals. …

Reasons why Intermittent Water Supply is not a satisfactory situation are presented in the book: it creates water quality problems, increases leakage and wastage, automatically reducing access to water, it creates unsolvable economic problems for the utility with detrimental effect on the structural integrity of the distribution network, substantial increase in the number of pipe bursts, customer dissatisfaction and reluctance to pay …

For the three billion people who still suffer from lack of water in quantity or accessibility, we need, as the UN is asking, to combine efforts of all, governments, civil society organizations and the private sector, in the context of a stronger and more effective global partnership for the development of appropriate policies and strategies in mitigating IWS.

This book is a first step in capacity building dealing with Intermittent Water Supply situations with the objective of improving practices. It is, I hope, the first of a long list of books and works on best practice in situations of IWS for helping professionals to cope better with it, still keeping the objective that one day, as soon as possible, everybody would have the possibility to have water 24 hours a day. But meanwhile we need to be able to be as intelligent as we can with the situations we have to face.

Read the book, share it, offer it, and discuss it. Welcome to the IWA network and if you can to the Intermittent Water Supply specialist group. … Thanks a lot to the authors!

Diane d'ARRAS
IWA President

Chapter 1

Introduction

Chrysi Laspidou, Alexandra Spyropoulou and Bambos Charalambous

Intermittent Water Supply (IWS) may be defined as piped water supply service that delivers water to users for less than 24 hours in one day. In an intermittent supply situation the consumers secure their water through the use of ground and/or roof tanks, where water is stored during the provided supply time. In a number of instances it could be that IWS is temporarily established to deal with extremely low water reserves due to prolonged drought or low rainfall periods resulting in temporary depletion of available water. In other instances IWS is a normal method of operation, such as in many countries in Asia, Africa and Latin America mainly due to the high levels of leakage in the networks, which cannot sustain continuous supply.

There is a lot of misconception surrounding IWS, the purpose it serves and the pressures it alleviates. Before getting into all the details and listing several arguments against IWS and presenting a useful guide for water specialists on how to deal with the critical issues leading to intermittent supply and how to transition to a 24 × 7 regime, a succinct list of reasons why IWS should be avoided is presented here. The list may shed light on issues that are often not readily acknowledged by many water professionals around the world; more data to support each reason presented here are included throughout the Chapters of this book.

1.1 10 REASONS WHY IWS SHOULD BE AVOIDED

(1) May seem to be a water saving measure however in the long run greater quantities of *water will be lost through increased leakage and wastage* compared to the quantities that may initially be saved.

(2) Has a *detrimental effect on the structural integrity of the distribution network* thus leading to quicker asset deterioration.

Bambos Charalambous and Chrysi Laspidou
doi: 10.2166/9781780407074_001

(3) Results in a *substantial increase in the number of pipe bursts* in mains and service connections thus increased leakage.
(4) *Creates water quality problems*, which are detrimental to human health and wellbeing and cause serious illness.
(5) Has an *adverse financial effect on the water utility* resulting in lower water sales and higher costs due to additional O&M activities needed to run IWS.
(6) Results in *customer dissatisfaction and reluctance to pay* due to poor quality of service provided.
(7) Entails an *additional financial burden* on the consumers.
(8) Is *not an effective Water Demand Management* measure.
(9) Is *not considered an appropriate intervention* to drought/water shortage.
(10) Takes "1 hour" to introduce but *years to reverse.*

1.2 DEFINITION: THE COMPLEXITY OF DEFINING AND MEASURING IWS

Intermittent Water Supply (IWS) refers to piped water supply service that is available to consumers less than 24 hours per day. In many regions around the world and mostly in Low And Middle Income (LAMI) Countries, IWS systems are prevalent; the situation is usually credited to shortage in water supply. An attempt to find a formal definition of IWS from the literature sheds light to the ambiguity that surrounds the issue and the lack of standardization in the terminology, which by itself leads to more confusion and makes taking action to address the problems that surround IWS more difficult. According to Galaitsi *et al.* (2016), terms referring to intermittent water supply include irregular, unreliable, inadequate and poor supply as well as outages and scarcity." Intermittency can be daily, seasonal or occasional and be related to inter-annual variability in water. In an IWS situation, consumers usually secure their water supply through the use of ground and/or roof tanks or smaller containers, where water is stored during the length of time that the supply is provided. Thus, it is possible that even though a city operates under an IWS regime, some consumers–especially wealthy – may not actually experience IWS, due to the fact that they have invested in pumps, tanks, filters, etc., so they store water when it is actually running and have enough supplies of purified water to use throughout the day. It is worth noting that IWS is enforced not only in cases where there is water shortage but also where the hydraulic capacity of distribution networks is such that cannot satisfy demand as well as in cases where the network is severely deteriorated resulting in high leakage (Charalambous, 2015).

Though basic issues related to water scarcity and intermittent supply are not part of popular consciousness in High Income (HI) countries, their effect is truly global as this issue affects over a billion people worldwide. Intermittent water supply systems are found mainly in LAMI countries, including most of the countries of South Asia and Latin America (Vairavamoorthy *et al.* 2008). In Latin America and the Caribbean, it is estimated that 60% of the population is served by household connections having intermittent service (PAHO and WHO, 2001). In Africa and Asia, it is estimated that more than one-third and one-half of urban water supplies, respectively, operate intermittently (WHO and UNICEF, 2000). According to WSP (2010), no city in India has continuous water supply and even in major cities of India, water supply is as low as 1 hour per day or less and 2–3 hours of supply is considered acceptable. Currently, some 30 LAMI countries worldwide are considered to be water stressed. It is predicted that by 2050, the number of water-scarce countries will likely approach 35 (Vairavamoorthy *et al.* 2007).

Intermittent distribution is widely adopted in both LAMI (McIntosh, 2003; McIntosh & Yñiguez, 1997; Hardoy *et al.* 2001) and HI countries (Cubillo, 2004, 2005) and it is usually done in order to distribute limited water quantities to as many people as possible. In this area, the lack of natural resources management and network maintenance plans, explicitly considering the possibility of scarcity scenarios, produces unexpected water shortage situations that can be handled only by mean of emergency interventions (Fontanazza *et al.* 2007).

'Full-pressure, "24 × 7" water supply remains a pipe dream in many cities' (World Bank, 2004). With the water demand increasing, especially in LAMI counties, often water utilities felt they have little option but to supply water intermittently. In many LAMI countries, intermittent water supply is still considered the norm (McIntosh, 2003).

1.3 WATER USE UNDER IWS AND 24 × 7 REGIME

When comparing water uses under the two different regimes, IWS and 24 × 7, the following question arises: to what extent do intermittent supplies reduce the overall demand for water? Consumers hoarding large amount of water and discarding it during the next flush simply results in excessive consumption, which contradicts the general belief that intermittent supply reduces overall water consumption. There are cases that document that water consumption under an IWS regime can be as much as double compared to the consumption when 24 × 7 conditions IWS (Kumpel & Nelson, 2013). Needless to say that this proves that per capita supply under intermittent supply is indeed sufficient to provide continuous service to consumers, even though lack of water supply is the most important reason cited for establishing an IWS scheme.

A succinct list of the benefits of 24 × 7 supply is the following:

24 × 7 supply delivers better quality water for public health: High levels of bacterial contamination are experienced in the first 10 minutes of re-pressurization of an intermittent system, in some cases persisting for up to 20 minutes. The contaminated water ends up in the consumers' storage tanks, which become also highly contaminated. Maintaining full pressure removes that risk. 24 × 7 supply gives significantly better service to all consumers.

24 × 7 supply revolutionizes service to the poor: Consumers can access more water for improved health and hygiene while saving time in queuing and carrying, and gainfully using the time thus saved for employment opportunities.

24 × 7 supply converts household coping costs into resources for the service provider: Coping costs that consumers need to incur are reduced; they pay for a better service.

24 × 7 supply reduces the burden on water resources: Continuous supply reduces water wastage arising from overflowing storage systems and open taps. It saves on stored household water that is discarded when new supply comes in. Because the network is renewed where needed, it also reduces losses arising from leaks in the old pipes.

24 × 7 supply delivers effective 'supply management' and 'demand management': Continuous supply makes possible the effective management of leakage through pressure management and flow measurement. Water conservation is also encouraged through metering and price signals via a volumetric tariff to consumers.

24 × 7 supply enables improved efficiency of service provision: Operational efficiencies are achieved because of a reduced need for valve-men, and a conversion of these jobs into more efficient ones of meter reading and customer care. It also makes possible the management of illegal connections (World Bank, 2004).

1.4 REFERENCES

Charalambous B. (2015). IWS Draft Concept Note for Book Publication, IWA.

Cubillo F. (2004). Droughts, risk management and reliability. *Water Science and Technology: Water Supply*, **4**(3), 1–11.

Cubillo F. (2005). Impact of end uses knowledge in demand strategic planning for Madrid. *Water Science and Technology: Water Supply*, **5**(3–4), 233–240.

Fontanazza C. M., Freni G. and La Loggia G. (2007). Analysis of intermittent supply systems in water scarcity conditions and evaluation of the resource distribution equity indices. *WIT Transactions on Ecology and the Environment*, **103**, 635–644.

Galaitsi S. E., Russell R., Bishara A., Durant J. L., Bogle J. and Huber-Lee A. (2016). Intermittent domestic water supply: a critical review and analysis of causal-consequential pathways. *Water*, **8**(7), 274.

Hardoy J. E., Mitlin D. and Satterthwaite D. (2001). Environmental Problems in a Urbanizing World: Finding Solutions for Cities in Africa, Asia and Latin America. Earthscan, London.

Kumpel E. and Nelson K. L. (2013). Comparing microbial water quality in an intermittent and continuous piped water supply. *Water Research*, **47**(14), 5176–5188.

McIntosh A. C. (2003). Asian Water Supplies Reaching the Urban Poor. Asian Development Bank.

McIntosh A. C. and Yñiguez C. E. (1997). Second Water Utilities Data Book: Asian and Pacific Region.

PAHO and WHO (2001). Regional Report on the Evaluation 2000 in the Region of the Americas. Pan American Health Organization, Washington, DC.

Vairavamoorthy K., Gorantiwar S. D. and Mohan S. (2007). Intermittent water supply under water scarcity situations. *Water International*, **2**(1), 121–132.

Vairavamoorthy K., Gorantiwar S. D. and Pathirana A. (2008). Managing urban water supplies in developing countries – climate change and water scarcity scenarios. *Physics and Chemistry of the Earth, Parts A/B/C*, **33**(5), 330–339.

WHO and UNICEF (2000). Global Water Supply and Sanitation Assessment 2000 Report. World Health Organization, Geneva.

World Bank (2004). World Development Report 2004. World Bank, Washington, DC.

WSP (2010). Water and Sanitation Programme. The Karnataka Urban Water Sector Improvement Project. 24 × 7 Water Supply is Achievable.

Chapter 2

Global dimensions of IWS – number of people affected worldwide

Chrysi Laspidou and Alexandra Spyropoulou

2.1 DISPARATE INFORMATION ON IWS

Most of the LAMI countries suffer from water scarcity; millions of people have no piped water on their premises. In many rural regions, in these countries, residents rely on a limited number of public taps (standpipe water), often operating intermittently and irregularly. Even for those residents with piped water on premises, water is supplied for limited hours per day or days per week. In the case of Katmandu, Nepal for example, residents receive water for one hour every other day during the dry season. In neighbouring India the average supply time countrywide is about 4.5 hours per day (Ministry of Urban Development, 2009). Population growth and urbanization rates are high in LAMI countries, making the financing of the expansion of water infrastructure where needed an increasingly difficult venture. In this backdrop, it is highly likely that IWS, which already is standard practice in many mega cities in the LAMI countries, is going to be more widespread (Vairavamoorthy *et al.* 2008). It is notable that more than one third of India's urban population has no access to safe drinking water and the majority of urban residents struggle to cope with both low pressure and IWS conditions.

IWS is a complex issue, encompasses a broad spectrum of practices with varying water delivery durations and patterns with social, economic and political implications. In order for the global community to appreciate the extent of these implications, it is important to know the number of people affected by IWS worldwide. Such estimation is attempted in this Chapter. It is interesting to note that even though IWS is a serious issue affecting a large part of the global population, there is lack of a systematic listing of countries and people affected by IWS. In general, there are incomplete data sets focusing on which countries receive IWS and for how many hours per day per week (Twort *et al.* 2000). A few published studies focus individually on a village or city, but there are no published data that systematically classify the water profile of each country.

When reviewing published data on IWS, an important issue that has become apparent is that different classifications of country groups are used by different researchers: For example, reporting numbers divided in country groups such as "East Asia & Pacific", "Europe & Central Asia" and "South Asia" can be confusing

Bambos Charalambous and Chrysi Laspidou
doi: 10.2166/9781780407074_005

since it is not always clear which countries are grouped under which country group. This way, data from the limited number of studies that are already published is often not comparable. Furthermore, the information about IWS is often contradictory since water supply and population are both dynamic in space and time and not all studies are based on the same census data, so the issue of outdated data is also important.

Kumpel and Nelson (2013) have recently published an article in which they provide an estimate of the global population affected by IWS (308.9 million). According to the literature, there are various discrepancies with their data. For instance, Butler and Memon (2006) estimates for South Asia show that at least 350 million people operate intermittently, compared to 116.6 million reported by Kumpel and Nelson (2013). Also, Choe *et al.* (1996) estimate that in Latin America more than fifty million inhabitants in ten of its major cities receive rationed supply, while Kumpel and Nelson (2013) report 28.4 million in Latin America, North America and the Caribbean together. Of course, it is possible that in the last couple of decades, there have been dramatic improvements that have led to the reduction of people under IWS regime, but according to experts' opinion in the field, it can safely be assumed that this is not the case.

2.2 PROPOSED METHODOLOGY FOR ESTIMATING IWS-AFFECTED POPULATION WORLDWIDE

The data reported in Kumpel and Nelson (2013) is based on the IBNET Water Supply and Sanitation Blue Book (Danilenko *et al.* 2014) from the latest year available for each utility, ranging from 1998–2014. The IBNET Blue Book presents a detailed statistical data on the operation of water utilities (including intermittency) for many LAMI countries. However, the list of utilities they include is not exhaustive, so they include only a sample of utilities. For instance, they only include data from 32 Indian utilities representing only 64 of the 370 million people. So, even though it is a great resource, the IBNET Blue Book cannot be taken as the representation of the "full picture", since they only represent a sample, which can be used to draw conclusions for the whole population of each country. Since the data reported in Kumpel and Nelson (2013) is based solely on the sampling data of the IBNET Blue Book, it was concluded that the population with IWS was underestimated and herein a different methodology to obtain a more realistic estimate closer to the real situation is presented.

However the utilities included in the IBNET database under each country do not necessarily constitute the total number of utilities existing and operating in each country. For instance in India 32 utilities are included in the database representing a total of 64 million people from a total of 370 million. Although the IBNET Blue Book is a great source of information and data on utilities listed therein the fact remains that not all the utilities in anyone county are included and therefore the presented data could not be representative of the country. Therefore since the data reported in Kumpel and Nelson (2013) is solely based on the data in the IBNET Blue Book, it is evident that the population under IWS is underestimated.

In order to obtain a more realistic estimate of the population and countries under IWS, a different methodology was used and is presented below:

Step 1. Define countries and country groups

In the present framework, the distinction in country groups as defined by the World Bank is used. This list includes six country groups, namely Africa, East Asia & Pacific, Europe & central Asia, Latin America & Caribbean, Middle East & North Africa and South Asia. Other classifications exist that are different to the above, but for consistency purposes, it was decided to follow the World Bank classification. In Table 2.1, the countries included under each of the above country group are listed; the 2-letter code of each country (used in charts and graphs in this Chapter) is also included. The analysis is on a country-by-country level, not on a country-group level, so it is important to specify in advance which countries are included in each country group. It is also important to recognize that in the World Bank classification, not all countries are included under each geographic region, but only the ones specified herein: For instance, under Europe & central Asia, countries of Western Europe are not included, but only specific ones that are relevant to this analysis. In the analysis, the countries listed in Table 2.1 only are considered, which is not to be confused with the name of the corresponding geographical region.

The analysis as presented in the Tables and Figures of this section, is based on the following numbers: The world population is 7.3 billion, out of which 4.2 billion has piped water on premises. The population in LAMI countries, as defined by the World Bank country group classification presented in Table 2.1, is estimated at 6.4 billion. Out of this population, 2.9 billion has piped water on premises and the analysis presented herein refers *only* to this population.

Step 2. Define Piped-on-Premises Population for all country groups

In the analysis, IWS is only defined for piped-on-premises populations. This excludes the population that has no access to a piped water network and relies either on standpipes, or on surface water. To collect the number of people with piped-on-premises facilities, the Joint Monitoring Programme (JMP) for water supply and sanitation by WHO and UNICEF (https://www.wssinfo.org/) was used, which gives the most updated information for water infrastructure worldwide. Based on this database, the following information was obtained separately for each country in each of the six country groups:

- National population, based on the latest census (year 2015 for most countries)
- Urban population
- Rural population
- Piped-on-premises population for each category (urban and rural)

Step 3. Classify each country as IWS or 24 × 7

After considering the lack of information from sources in the published literature, it was decided to make realistic assumptions in order to estimate populations affected by IWS. The IBNET Blue Book (Danilenko *et al.* 2014) was used to decide whether a country has IWS or not. As mentioned earlier, the IBNET Blue Book provides data for only a number of utilities in each country. Based on this information it was considered that it would be reasonable to assume that if the listed utilities experience IWS conditions it is highly probable that similar conditions will be experienced countrywide. Thus, when continuity of service is listed at less than 24 hours in the IBNET Blue Book, the specific country was classified as IWS and included all the piped-on-premises population for that country (Step 2) as IWS population. In several cases, for which there were no published data, a separate search for each country was conducted to find any relevant publication, either scientific, or news articles, or an Internet source in order to identify specifics on the continuity of water supply for that country. Additionally people in the IWA network that have knowledge on specific countries as well as individuals from countries were conducted/interviewed. Based on all the information an IWS or 24 × 7 classification for countries was established. It is recognised that with this methodology, the total number of people affected by IWS may have been overestimated, but all possible effort was made to get a reasonable estimate.

Step 4. Sum-up and obtain statistics

Once all data was gathered, sums and percentages for each country in each one of the six country groups were created and presented in this Chapter.

Table 2.1 Country group classification and corresponding countries in each group.

Africa	Angola (AO), Benin (BJ), Botswana (BW), Burkina Faso (BF), Burundi (BI), Cameroon (CM), Cabo Verde (CV), Central Africa Republic (CF), Chad (TD), Comoros (KM), Cote d'Ivoire (CI), Eritrea (ER), Ethiopia (ET), Gabon (GA), Gambia (GM), Ghana (GH), Equatorial Guinea (GQ), Guinea (GN), Guinea-Bissau (GW), Kenya (GW), Lesotho (LS), Liberia (LR), Madagascar (MG), Malawi (MW), Mali (ML), Mauritania (MR), Mauritius (MU), Mozambique (MZ), Namibia (NA), Niger (NE), Nigeria (NG), Rwanda (RW), Sao Tome and Principe (ST), Senegal (SN), Seychelles (SC), Sierra Leone (SL), Somalia (SO), South Africa (ZA), South Sudan (SS), Sudan (SD), Swaziland (SZ), Tanzania (TZ), Togo (TG), Uganda (UG), Zambia (ZM), Zimbabwe (ZW).
East Asia & Pacific	China (CN), Cambodia (KH), Indonesia (ID), Japan (JP), Korea, Lao PDR (LH), Malaysia (MY), Myanmar (MM), Pacific Islands, Papua New Guinea (PG), Philippines (PH), Singapore (SG), Thailand (TH), Timor Leste (TL), Vietnam (VN), Mongolia (MN).
Europe & Central Asia	Albania (AL), Armenia (AM), Azerbaijan (AZ), Belarus (BY), Bosnia and Herzegovina (BA), Bulgaria (BG), Kyrgyz Republic (KG), Tajikistan (TJ), Uzbekistan (UZ), Turkmenistan (TM), Ukraine (UA), Croatia (HR), Georgia (GE), Kazakhstan (KZ), Kosovo, FYR of Macedonia (MK), Moldova (MD), Montenegro (ME), Poland (PL), Romania (RO), Russia Federation (RU), Turkey (TR), Serbia (RS).
Latin America & Caribbean	Antigua and Barbuda (AG), Argentina (AR), Belize (BZ), Bolivia (BO), Brazil (BR), Chile (CL), Colombia (CO), Costa Rica (CR), Dominica (DM), Dominican Republic (DO), Ecuador (EC), El Salvador (SV), Grenada (GD), Guatemala (GT), Guyana (GY), Haiti (HT), Honduras (HN), Jamaica (JM), Mexico (MX), Nicaragua (NI), Panama (PA), Paraguay (PY), Peru (PE), St. Kitts and Nevis (KN), St. Vincent and Grenadines (VC), St. Lucia (LC), Suriname (SR), Uruguay (UY), Venezuela (VE).
Middle East & North Africa	Algeria (DZ), Bahrain (BH), Djibouti (DJ), Egypt (EG), Iran (IR), Israel (IL), Iraq (IQ), Jordan (JO), Kuwait, Lebanon (LB), Libya (LY), Morocco (MA), Oman (OM), Qatar (QA), Saudi Arabia (SA), Syria (SY), Tunisia (TN), United Arab Emirates (AE), West Bank and Gaza, Yemen (YE).
South Asia	Afghanistan (AF), Bangladesh (BD), Bhutan (BM), India (IN), Maldives (MV), Nepal (NP), Pakistan (PK), Sri Lanka (LK).

2.3 RESULTS

In 2015, it is estimated that 663 million people worldwide still use unimproved drinking water sources (https://www.wssinfo.org/), including unprotected wells, springs and surface water (UNICEF and World Health Organization, 2015). The percentage of people lacking *piped water supply* is higher in rural regions in LAMI countries. It should be noted that under the context "rural regions" in this Chapter, *only* the rural regions which have *piped water on premises* are included. Throughout the last 25 years (from, 1990 to 2015), more improvements in water infrastructure have been realized in urban rather than in rural areas, probably due to the fact that large numbers of people have moved to urban areas. Although in total (worldwide), there is improvement in extending piped water infrastructures, the progress in rural regions is slower.

According to the 2015 Census, the population worldwide is 7.3 billion of which 88% (6.4 billion) lives in LAMI countries, the majority of these are in the most water-stressed areas of the world. Only 57% of the world population (4.2 billion) has access to piped water on premises, Figure 2.3, of which 74% (3.1 billion) lives in urban areas, Figure 2.1, whereas 26% (1.1 billion) lives in rural areas, Figure 2.2.

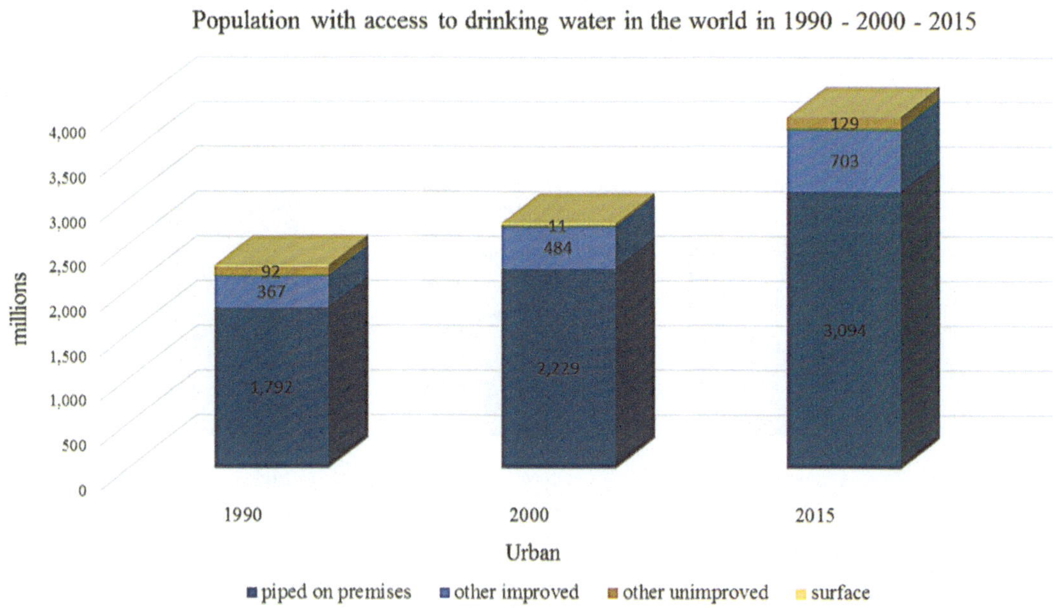

Figure 2.1 Worldwide *urban* population with access to piped water. (Graphs produced from data on the following website: http://www.wssinfo.org/data-estimates/graphs/).

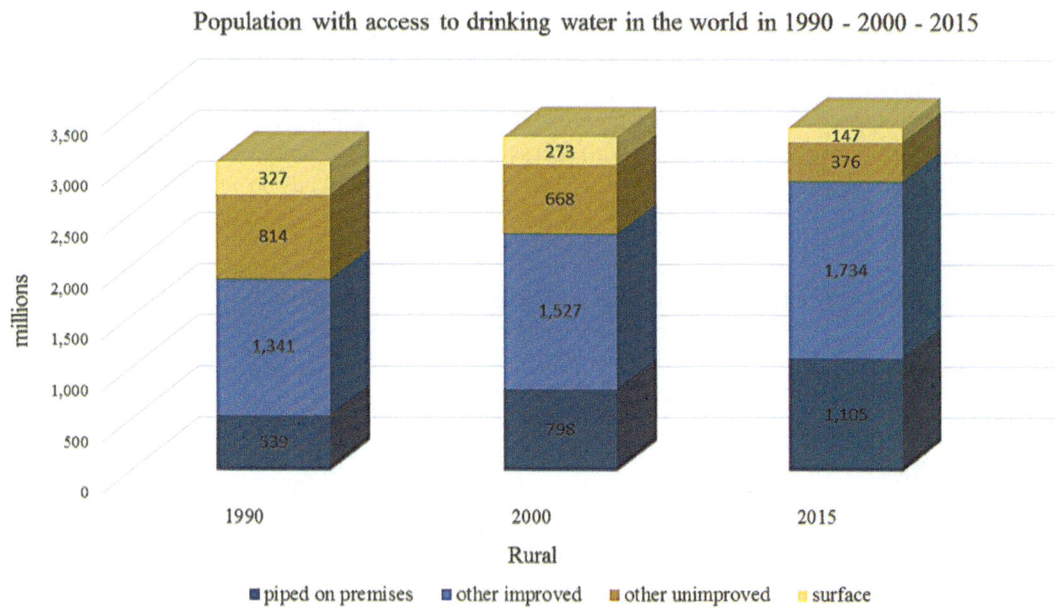

Figure 2.2 Worldwide *rural* population with access to piped water. (Graphs produced from data on the following website: http://www.wssinfo.org/data-estimates/graphs/).

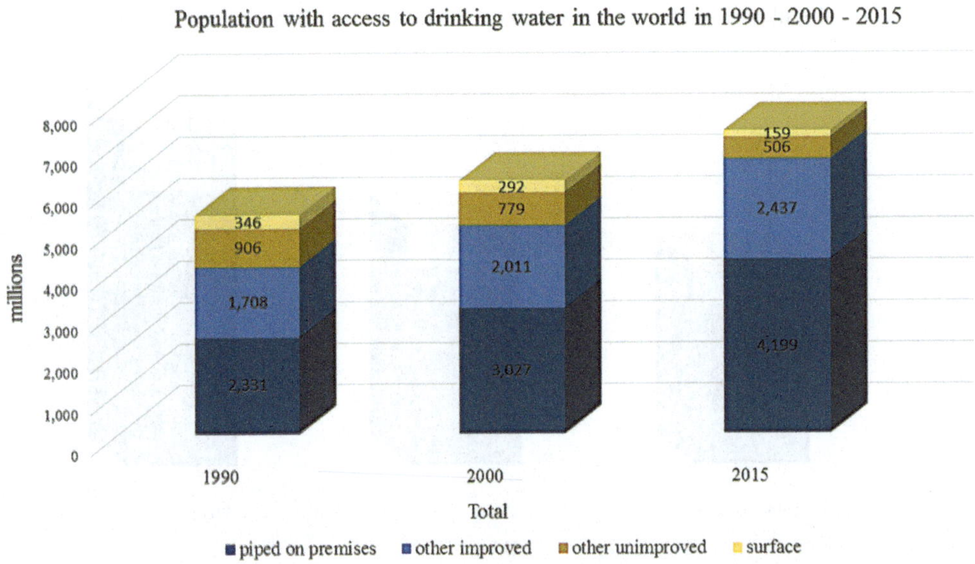

Figure 2.3 Worldwide population with access to piped water. (Graphs produced from data on the following website: http://www.wssinfo.org/data-estimates/graphs/).

In order to have a better picture of the population with piped water on premises in LAMI countries, the number of the population with piped water on premises to the total population was compared. The population of each country group, which has piped water on premises, is shown in Table 2.3. Comparing the data from Table 2.2 to Table 2.3, it can be seen that Middle East & North Africa Group and Latin America & Caribbean Group have the highest percentage (88% and 91% respectively) of people with access to piped water whereas Europe & Central Asia and Africa have the lowest percentage (16% and 18%). The urban, rural and total population with access to piped water is shown graphically in Figure 2.4. It is evident that a large part of the population in each country group still has no access to piped water on premises.

Table 2.2 Urban, rural and total population in LAMI countries.

	Population (Million)		
LAMI Country Groups	**Urban**	**Rural**	**Total**
Africa	370	608	978
East Asia & Pacific	1,243	978	2,221
Europe & Central Asia	290	160	450
Middle East & North Africa	274	164	438
Latin America & Caribbean	489	123	612
South Asia	566	1,148	1,714
Totals	3,232	3,181	6,413

Data source: WHO/UNICEF JMP for water supply and sanitation.

Table 2.3 Urban, rural and total population with piped water on premises in LAMI countries.

	Population with Piped Water on Premises (Million)		
LAMI Country Group	**Urban**	**Rural**	**Total**
Africa	147	36	176
East Asia & Pacific	843	409	1,252
Europe & Central Asia	53	19	72
Middle East & North Africa	249	135	383
Latin America & Caribbean	469	86	555
South Asia	297	176	472
Totals	2,050	860	2,910

Data source: WHO/UNICEF JMP for water supply and sanitation.

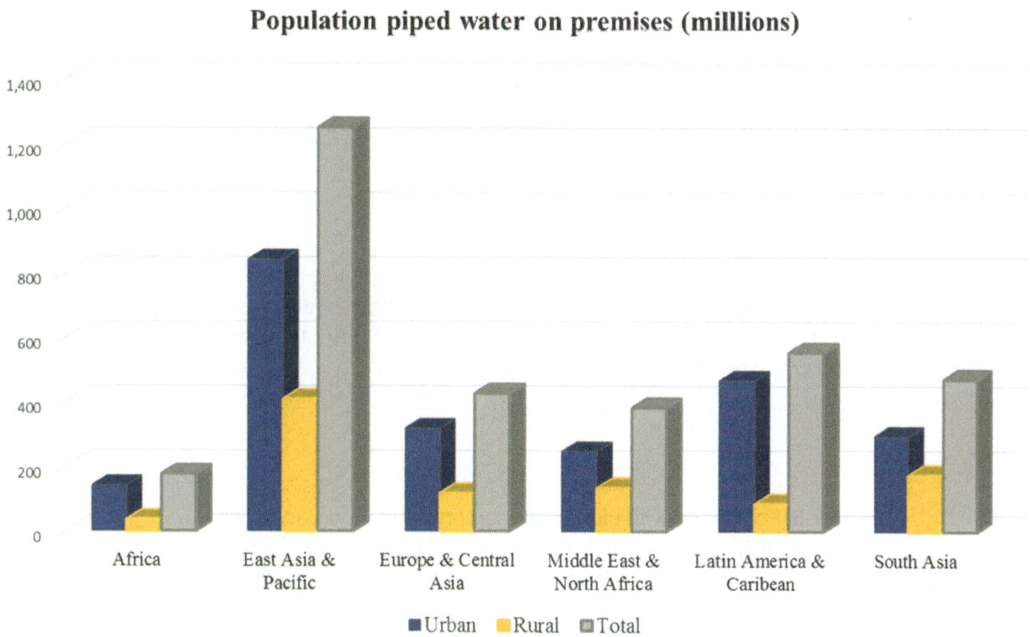

Figure 2.4 Urban, rural and total population with access to piped water for LAMI country groups.

Furthermore, as expected, it is observed that for all country groups in rural areas the population with piped water on premises is less, compared to the urban areas/small cities. In Africa, it is remarkable that only 6% of the rural population has access to piped water, in Europe & Central Asia 12%, in South Asia

15%, in East Asia & the Pacific 42% and 70% and 82% in Latin America & Caribbean and Middle East and North Africa respectively.

When it comes to IWS, the analysis revealed that in Africa, 48% of the population with piped water on premises is affected by IWS, in East Asia & the Pacific 13%, in Europe & Central Asia 100%, in Middle East & North Africa 65%, in Latin America & the Caribbean 48% and finally in South Asia 100%. As it is observed, the percentages are very high, with all country groups being higher that 48%, with the exception in East Asia & the Pacific. It is worth noting that in Europe & Central Asia as well as in South Asia almost all the population is affected by IWS.

Table 2.4 Population with piped water affected by IWS in LAMI countries.

	Population Affected by IWS (Million)		
LAMI Country Group	**Urban**	**Rural**	**Total**
Africa	67	17	84
East Asia & Pacific	121	48	169
Europe & Central Asia	53	19	72
Middle East & North Africa	187	62	249
Latin America & Caribbean	216	50	266
South Asia	297	175	472
Totals	941	372	1,313

In summary the analysis revealed that from a total of 3,195 million people with access to piped water on premises in LAMI countries, 1,313 million are affected by IWS of which 941 million are located in urban areas and 372 million in rural areas (Table 2.5). IWS affects approximately 41% of the population with piped water on premises (Figure 2.5).

Table 2.5 Population with piped water on premises and population affected by IWS for urban and rural areas and in total for LAMI countries.

	Population (Million)		
	Urban	**Rural**	**Total**
LAMI countries population	3,232	3,181	6,413
Piped water on premises	2,050	935	3195
Population affected by IWS	941	372	1,313

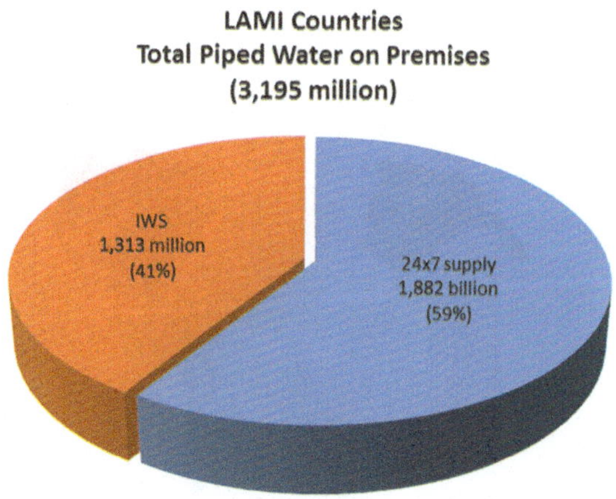

Figure 2.5 Total population with piped water on premises, considering all country groups and population affected by IWS for LAMI countries.

Table 2.6 Worldwide population, population with piped water on premises and population affected by IWS calculated only for the LAMI countries (urban, rural and total).

	Population (Million)		
Description	**Urban**	**Rural**	**Total**
Worldwide Population	3,939	3,362	7,301
Population in HM countries with piped water on premises	707	181	888
Population in LAMI countries	3,232	3,181	6,413
Population with piped water on premises in LAMI countries	2,260	935	3,195
Population with piped water on premises affected by IWS in LAMI countries	941	372	1,313

Data source: WHO/UNICEF JMP for water supply and sanitation.

A summary of the global figures giving an estimate of the population with piped water on premises that is affected by IWS is presented in Table 2.6: Out of all urban population worldwide (3,939 million), 75% (2,967 million) has piped water on premises, of which 32% (941 million) is affected by IWS and is located principally in LAMI countries. Out of all rural population worldwide (3,362 million), only 33% (1,116 million) has piped water on premises, of which 33% (372 million)is affected by IWS located primarily in LAMI countries. From the worldwide population (7,301 million), only 56% (4,083 million) has piped water on premises, of which 32% (1,313 million) is affected by IWS principally in LAMI countries.

In the Figures that follow (Figure 2.6 to Figure 2.9), maps of the country groups considered are presented showing countries with 24 × 7 water supply and population in countries affected by IWS.

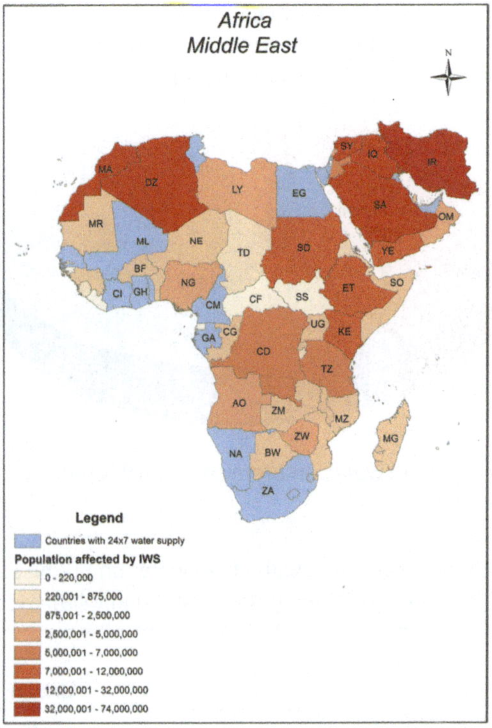

Figure 2.6 Map of Africa and Middle East countries showing 24 × 7 water supply and population affected by IWS.

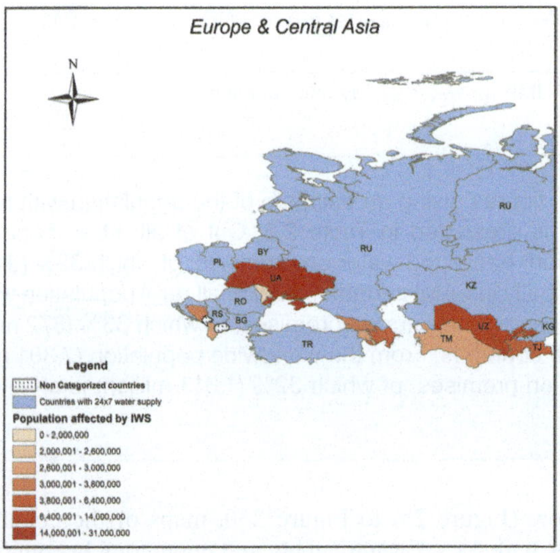

Figure 2.7 Map of Europe and Central Asia countries showing 24 × 7 water supply and population affected by IWS.

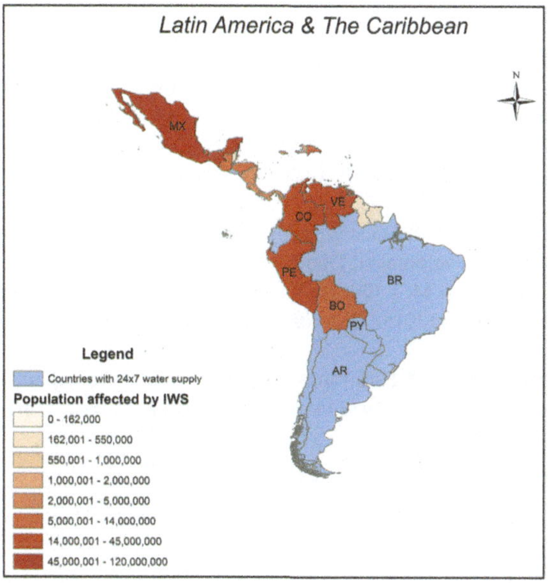

Figure 2.8 Map of Latin America and Caribbean countries showing 24 × 7 water supply and population affected by IWS.

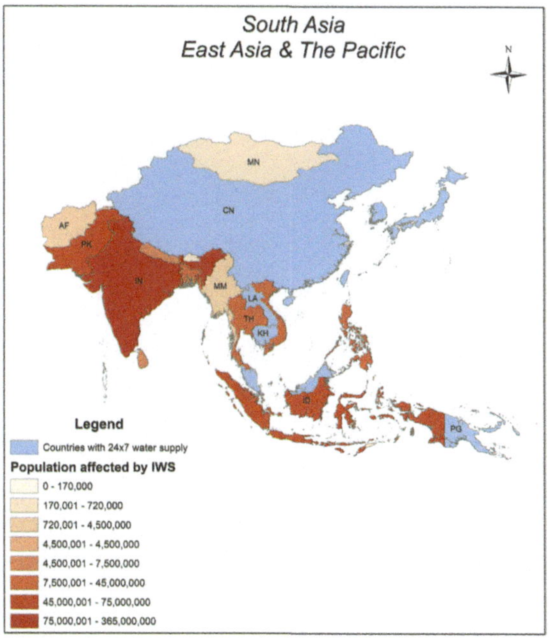

Figure 2.9 Map of South East Asia and Pacific countries showing 24 × 7 water supply and population affected by IWS

2.4 REFERENCES

Butler D. and Memon F. (2006). Water Demand Management. IWA Publishing, London, UK.

Choe K., Varley R. C. G. and Bijlani H. U. (1996). Coping with Intermittent Water Supply: Problems and Prospects, Dehra Dun, Utter Pradesh, India (Activitiy Report No. 26). Environment Health Project, Washington, DC.

Danilenko A., Van den Berg C., Macheve B. and Moffitt L. J. (2014). The IBNET Water Supply and Sanitation Blue Book 2014: The International Benchmarking Network for Water and Sanitation Utilities Databook. World Bank Publications.

Ministry of Urban Development (2009). Service Level Benchmarking Databook: Improving Service Outcomes 2008–2009. Ministry of Urban Development, Delhi.

Twort A. C., Ratnayaka D. D. and Brandt M. J. (2000). Water Supply. Butterworth-Heinemann.

UNICEF and World Health Organization (2015). Progress on Sanitation and Drinking Water–2015 Update and MDG Assessment, WHO Press, Geneva.

Vairavamoorthy K., Gorantiwar S. D. and Pathirana A. (2008). Managing urban water supplies in developing countries–climate change and water scarcity scenarios. *Physics and Chemistry of the Earth, Parts A/B/C*, **33**(5), 330–339.

Chapter 3

Root causes and implications of IWS

Chrysi Laspidou, Alexandra Spyropoulou, Bambos Charalambous and Shrihari Sridhar

3.1 ROOT CAUSES

When it comes to the root causes of IWS, one thing is for certain: The first step in solving a problem is recognizing one exists. In this Chapter, all crucial factors that lead to, or are said to justify, intermittent supply are presented. The common understanding on the reason behind intermittency is that the amount of water available is insufficient to meet consumer demands. But a little more attention would reveal that intermittency persists even in places, which have abundant fresh water resource. For instance, even though the Indo-Gangetic plains, home to a billion people in the Indian subcontinent (which covers the Ganges-Brahmaputra region of Eastern India and Bangladesh) have perennial water source in abundance, no city in the region enjoys a continuous water supply. Such realities prove that apart from absolute water scarcity there are other factors, which are either individually or collectively responsible for the intermittency.

This position can also be strengthened by the following comparison: The Water Stress Index (WSI) released by risk analysis and mapping firm Maplecroft (2010), pinpoints areas of water stress down to 10 km² worldwide by calculating the ratio of domestic, industrial and agricultural water consumption, against renewable supplies of water from precipitation, rivers and groundwater. Combining data from the IBNET Blue Book (Danilenko *et al.* 2014), which classifies countries with IWS regime in four categories according to the hours of water supply per day, and the WSI as released by Maplecroft, a link between the hours of supply per day (extent of IWS) to Water Stress Index (WSI) was established as shown in Table 3.1. It is expected that the countries with extreme WSI would classify mostly under the "<10 hours/day" category in terms of IWS, countries with high WSI under the "10–15 hours/day" and so on. However, it can be seen that in the "<10 hours/day" category, all countries listed appear to have a low stress WSI (even if it is partly for India). The Gulf nations are rated as the world's most water stressed countries, with the least available water per capita, by a new ranking of 186 countries (Maplecroft, 2010). Countries such as

doi: 10.2166/9781780407074_017

Bahrain, Qatar, Kuwait and Saudi Arabia with extreme WSIs have no IWS at all, thus disproving the myth that IWS is all about water scarcity.

A combination of the factors is suggested as the root cause of IWS (Charalambous, 2015) including water scarcity, fast population growth in LAMI countries, rapidly increasing demand due to urbanization, alternatives are perceived as impossible, lack of planning – poor management and lack of awareness. In the sections that follow, each one of these root causes are examined separately. International experience has shown that in most cases leakage is the primary cause of IWS, which forces operators to manage the networks intermittently. In fact, intermittency may start with a few hours a day but as the leakage situation becomes worse, the hours of supply become less and less in order to avoid the loss of vast quantities of water lost through the leaks. It has been well established and will be presented in later chapters in this book that IWS causes detrimental effects to the structural integrity of the network. More breaks appear in the network as time passes, due to the vicious cycle of emptying and refilling the network, thus subjecting the pipes daily to a huge pressure variation and reversal of stresses and strains reducing drastically their factors of safety. The link between water losses and IWS is explored in Chapter 5. In the sections that follow, an analysis into each one of the aforementioned root causes of IWS is carried out.

Table 3.1 IWS around the world (hours/day data originates from Danilenko *et al.* 2014).

<10 Hours/Day	<10–15 Hours/Day	<15–20 Hours/Day	<20–24 Hours/Day
Nigeria (LS)	Pakistan (ES)	Mozambique (LS)	Tajikistan (ES)
India (LS, MS, HS, ES)	DR Congo (LS)	Zimbabwe (ES)	Kyrgyzstan (HS, ES)
Guinea (LS)	Ghana (LS)	Tanzania (LS, ES)	Ukraine (LS, HS)
	Armenia (LS)	Kenya (ES)	Romania (LS, MS, ES)
	Georgia (LS)	Sudan (ES, ND)	Bulgaria (LS)
	Albania (LS)	Sri Lanka (LS, HS, ES)	FYROM (LS, HS)
		Indonesia (LS, MS)	Cyprus (ND)
		Philippines (LS, ND)	Ethiopia (ES, ND)
		Venezuela (LS, HS, ES)	Uganda (ND)
		Peru (HS, ES)	Vietnam (LS, MS)
		Bolivia (LS, ND)	Mexico (LS, MS, HS, ES)
		Panama (LS, ND)	
		Nicaragua (LS, HS, ES)	
		Uzbekistan (LS, HS)	

Note: ES refers to Extreme Stress; HS refers to High Stress; MS refers to Medium Stress; LS refers to Low Stress; ND refers to No Data.

3.1.1 Water scarcity

Water scarcity is cited as one of the most important root causes for IWS. Water scarcity refers to the problematic relationship between water availability and human population, i.e. low water availability per capita per year, usually on a national scale (Gleick, 1996). According to Rijsberman (2006), when an individual does not have access to safe and affordable water to satisfy her or his needs for drinking, washing or their livelihoods that person is called water insecure. When a large number of people in an area are water insecure for a significant period of time, then the area could be called water scarce. It is important to note, however, that there is no commonly accepted definition of water scarcity. Whether an area qualifies as "water scarce" depends on, for instance: (a) how people's needs are defined and whether the needs of the environment, the water for nature, are taken into account in that definition; (b) what fraction of the resource is made available, or could be made available, to satisfy these needs; (c) the temporal and spatial scales used to define scarcity. Three water scarcity categories established

- Scarcity from poor management (high leakage & wastage, poor O&M practices, governance issues – corruption, etc.)
- Economic scarcity (financial constraints to expand infrastructure to meet increasing demand; poor planning & demand forecasting)
- *Absolute scarcity (insufficient quantity at the source – exacerbated by climate change) (Source*: Harrison Mutikanga)

Lack of access to a continuous source of clean water is a central problem in LAMI countries. A recent global estimate of water supply in the LAMI world suggests that 884 million people still require access to safe drinking water (WHO/UNICEF Joint Water Supply, Sanitation Monitoring Programme, World Health Organization, & UNICEF, 2008). Optimistically, this indicates that 1.6 billion people have now been granted access to safe drinking water since the 1990's. Some of these solutions have been large, centralized and capital intensive projects like piped water systems and dams. In Sub-Saharan Africa, one of the poorest areas in the world, where generally speaking these large scale solutions remain unavailable many of the inhabitants in this area are classified globally, as the "bottom billion" (Collier, 2012). Also, more than a third of India's urban population has no access to safe drinking water, and the majority of urban residents struggle to cope with both low and intermittent water supply conditions. By the beginning of the 21st century, the country's urban population will have expanded by one-third, to 320 million, further exacerbating India's water supply problem. The importance of water security for all is reflected in the United Nations Sustainable Development Goal no. 6, which sets as a goal to ensure availability and sustainable management of water and sanitation for all.

3.1.2 Fast population growth in LAMI countries

The most significant reason behind inadequacy of domestic water supply – a root cause for IWS – is the explosive growth of demand. There are two important aspects of this increase: first, it is the absolute increase of demand due to population growth, but also the growth of per-capita water use in parallel to the economic growth and improvement of living conditions. The second aspect is the increasing concentration of population in urban centres. Increasing population densities creates huge demands concentrated in urban centres demanding extensive measures in conveying water from remote locations and distribution (Vairavamoorthy *et al.* 2008). Over the next few years, an increasing population and increasing use of water will put increasing pressure on global water resources: pressures will increase most rapidly in Africa and parts of Southern Asia (McIntosh, 2003).

The rapid population growth that has been documented in the years from 1970 to 1994 has resulted in the potential water availability for the Earth's population decreasing from 12.9 down to 7.6 thousand cubic metres per year per person. The greatest reduction in population water supply took place in Africa (by 2.8 times), in Asia (by two times), and in South America (by 1.7 times)(Shiklomanov, 1991).

Demographic pressure has led one international water resources experts to predict that lack of adequate supplies of renewable fresh water could soon become the main constraint on their economic development (Biswas, 1992). This is already the case in several water scarce countries in northern and Southern Africa, where population growth rates remain high and increased pressure is placed on limited water resources to meet the growing demand for food production. Many countries that will cross the water scarcity benchmark within the next 10 years are in Africa thus adding more than 100 million people to those already coping with severe water resource shortages (Gardner-Outlaw & Engelman 1997). In these and many other countries, increasing water scarcity will increase the cost of producing and delivering fresh water just to keep pace with current levels of service. This increased cost will drain financial resources that would otherwise be available for investment in the development of other sectors of the economy (Khroda, 1996). One can only imagine how much these projected changes will worsen the situation for countries currently operating under IWS regime – this is one more reason why understanding IWS and offering solutions that will facilitate the transition to 24 × 7 service becomes increasingly important.

3.1.3 Rapidly increasing demand due to urbanization

Migration of the rural population to urban centers has resulted in towns and cities expanding rapidly. For example between 1950 and 1990 the number of cities with populations of more than 1 million increased from 78 to 290 and this number is expected to exceed 600 by 2025 (Serageldin, 1995). On a global scale, half of the world's population lives in urban areas today. Out of this urban population, two out of three are in LAMI countries. By year 2025, there will be four urban dwellers in LAMI countries for each one in the rest of the world. The increasing concentration of populations in urban areas and the growth of mega cities place a strain on existing public services and, together with prevailing economic realities, result in chaotic conditions in many towns and cities throughout the LAMI world. According to a study by WHO and UNICEF, the number of urban residents without access to improved water sources in fact rose from 113 million in 1990 (5% of the total urban population) to 173 million in 2000 (6% of the total urban population) (Hinrichsen *et al.* 2002). Water supply to cities is crucial not only for the survival of its inhabitants but also to maintain a healthy level of economic development. Awareness of what happens when there is a lack of water is all too apparent in many third world cities, where poor living conditions and limited access to adequate water and sanitation significantly increase the health burden on the urban poor, who often constitute the very labour source that generates the cities' wealth (UNESCO, 2003). For example, 85% of India's urban population has access to drinking water but only 20% of that water meets the minimum health and safety standards. It is estimated that by the year 2050, a half of India's population will be living in urban areas and will face acute water problems (Singh, 2000). Naturally, this rapidly increasing water demand due to urbanization only makes matters network rehabilitation, reduction of physical losses and conversion to 24 × 7 a total necessity.

3.1.4 Alternatives are perceived as impossible

For many countries, IWS has been a reality for so long, so people do not see the urgent need for change. In South Asia, the use of IWS is so high that most people regard it as normal and are therefore not greatly motivated to do anything about it (McIntosh, 2003). In Kathmandu, in the dry season, most people get water for about one hour every other day. In Indian cities, two or three hours of water a day is considered good (McIntosh, 2003).

IWS is usually adopted through necessity in water stressed regions (Vairavamoorthy *et al.* 2008) and is very common in LAMI countries, especially in South Asia. Changes in water supply can be effected by Asia's growing middle class and its small but very powerful wealthy group, if they decide to use their influence; however, motivation for such decision is usually low, since they secure their water supply as individuals through the use of tanks, pumps, and in many cases private wells (McIntosh, 2003). As long as the interests of the wealthy are served, the motivation for changes will be weak.

Faced with the pressure of scarcity, water authorities will increasingly resort to delivering intermittent supplies, if they are not convinced that alternatives do exist. Usually during drought periods, water utilities impose water restrictions to both domestic and agricultural supplies. However, in most cases, water authorities seem to overlook the obvious, which is to manage the water networks in the most efficient and effective way in order to minimize physical losses through leakage. Some of the alternatives, that could be realized with careful financial and governmental planning, include the construction of treatment units to treat domestic effluent for agriculture, or the construction of desalination plants to produce potable water for satisfying domestic needs, thus adding to the water balance and reducing deficit.

3.1.5 Lack of planning

It turns out to be difficult to assess whether water is truly scarce in the physical sense (a supply problem) or whether it is available but should be used better (Rijsberman, 2006). There is much talk of a water crisis, of which the most obvious manifestation is that 1.2 billion people lack access to safe and affordable water for their domestic use (WHO, 2003). IWS is adopted due to water scarcity with the known results in inequities in water deliveries to consumers and poor levels of service. It is therefore important to recognise these realities when designing and operating such networks and to examine the role that proper planning might play in avoiding to adopt such a scheme (Vairavamoorthy *et al.* 2007).

The lack of organizing the use of IWS drives to inefficient and ineffective results. The creation and maintenance of a plan is necessary in order to achieve optimum balance of needs or demands with the available resource. In many cases, lack of planning drives to wasteful use of existing water supplies (Rosegrant *et al.* 2002). In a planning process under IWS, many aspects must be considered.

The knowledge of the current situation (the quality of available level data is far from optimal) is vital. The gaps and limitations of data available from public sources and the lack of regular monitoring represent serious constraints to the understanding of current and future needs and to the ability to plan for meeting these needs. There is not enough data available for almost all areas relevant to urban water management – particularly for informal settlements – including data on hydrometeorology, water resources, water services, and coping strategies, as well as information on related factors such as population and land use (Jacobsen *et al.* 2013).

Another important issue is which hours are better to have water and for what uses water is spent. Water is required for agricultural, industrial, household, recreational and environmental uses. The demand for these uses is rapidly increasing due to continuously increasing population especially in the LAMI world and due to the growing awareness of environmental, health and recreational issues. At the same time, developing a plan that prioritises water uses and expands water availability through new infrastructure becomes more difficult as the cost of developing new sources or expanding existing ones is continuously increasing as the most accessible water resources have already been tapped (UNESCO, 2003).

The equality of water allocation between the poor and the rich is still an open issue in LAMI countries. Water policies and institutions must be designed to improve access for the poor, women, landless, and underprivileged social groups to water for irrigation and household uses, livestock, fishing and micro-enterprises (Gottipati & Nanduri, 2014). At most of the LAMI countries facing the biggest scarcity problems, governments lack the expertise on how to develop a plan or to maintenance it (Collier & Venables, 2012). In addition, there is lack of experienced and trained human recourses. The right people need to accomplish the plan and transfer the knowledge from other regions.

A major reason for the poor performance of intermittent supply systems is the fact that they are often adopted in a reactive way as opposed to proactive planning. Today it is important to recognize intermittent supplies as a reality that is eventually hard to escape and an attempt must be made to be proactive in the design of IWS management plans at least as fallback plans to ensure adequate service standards (Khroda, 1996).

3.1.6 Lack of awareness

People are generally not actively opposed to the use of intermittent supply, mostly because they are unaware of the consequences. Lack of awareness, education and understanding of the subject drives public to consider IWS as the norm, without knowing that 24-hour supply to the home is the norm in most countries in the world. Consumers are often unaware of government policies and there is low government accountability; as a result, poorly informed consumers do not demand for better services and do not exert pressure to the government.

The public is not aware about the potential problems of intermittent supply, such as health risks, water contamination, wasting water, meter malfunctioning, overexploitation of groundwater and incurring high costs in consumers and providers tied to intermittent supply.

Public education and awareness are important. In today's digital societies, it is possible to keep many consumers informed of such matters (tariffs, coping costs). And those without access to Internet can be informed through radios, televisions, and newspapers. When this happens, the consumers will be more in control. More public awareness, public relations, and public involvement programs are needed, as these will ensure the public can learn the facts about IWS. Everyone involved, civil society (represented by NGOs, journalists, and academics) needs to be aware, educated and more motivated of the problems tied with IWS. Then it will be necessary to embark on extensive stakeholder awareness programs to convince people that 24-hour access to piped water in the home is possible for all. It is important to build awareness and create the environment to make informed choices (McIntosh, 2003).

3.2 IMPLICATIONS OF IWS

The implications of the intermittent supply are many and complex. The most obvious practical consequences of having IWS are: water contamination and public health issues, increased mains failure and network deterioration, increased difficulties in detecting and fixing leaks, illegal connections and/or meter tampering, meter malfunctioning and accelerated wear and tear, ineffective supply and demand management and inefficient operations, which require more manpower. The increasing demand of water leads to a reduction in supply and lower supplies eventually means that it is not possible to supply 24 × 7. As a result, IWS is used and pressure strongly fluctuates in the distribution network, which has been originally designed to operate under constant pressure. This leads to repeated network failures and pipe bursts, a deteriorating network, and increasing water losses due to leakage, while Operation and Maintenance (O&M) costs keep getting higher. Unable to rectify the situation created, leakages continue to increase leading to further reduction of the water supply. At that point, the vicious cycle of intermittent supply starts, which, even though was initially implemented to deal with short supplies, invariably ends up exacerbating the cause.

IWS can be seen as a downward spiral (schematically shown in Figure 3.1): Increased urbanization leads to higher water demand. Water companies tend to respond with network expansion, which often takes place after poor planning and extends the network beyond capacity, thus lowering quality of service for consumers. This leads to an inadequate water supply for towns and agglomerations, which may lead people (mostly the privileged) to take matters in their own hands and proceed with private investments that will improve their service. The result is a poor service for most people, which inevitably leads to corruption and conflict among different citizen groups. Such instabilities lead to less willingness to pay for water by the public, which in turn leads to less income for water utilities, which see more bursts in the network, need more staff to respond to the failures and are faced with lack the funds due to reduced income. The result is a deteriorating network with increased leakage and higher Non-Revenue Water (NRW) overall. At the end of the day, more money is being spent on a system that provides lower quality service to the public, which leads to the realization of the following paradox: Intermittency may be implemented when funds are limited, but ends up being an expensive regime to follow!

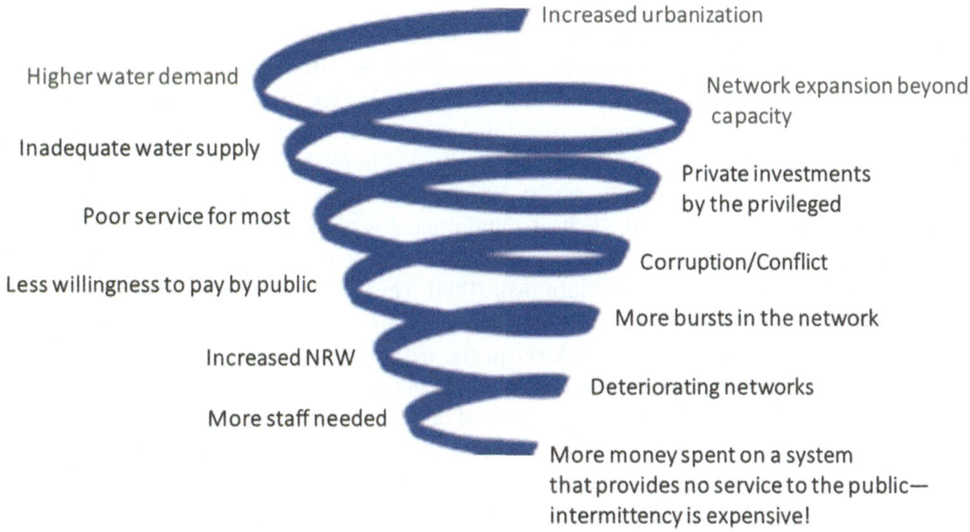

Figure 3.1 The downward spiral of IWS.

The serious implications of IWS do not stop there. One cannot ignore that IWS is a deeply political problem that creates serious socio-economic inequalities and is often unrelated to the simple explanations offered by most. IWS creates a gender issue since it affects young girls and women the more – typically it is those who are in charge of getting the water in the household. IWS is a social issue, as it affects mostly the poor and disadvantaged of a society; the privileged usually take matters in their own hands and invest in private solutions to get the water they need. As a result, IWS extends corruption and conflict within communities and deepens the gap between rich and poor. IWS affects productivity, since it is not uncommon to have one person in the family fully assigned to the task of securing the water supply. It also disrupts everyday life, since often the daily schedule of a household is adjusted around the water supply schedule, which is sometimes irregular and unpredictable. Finally, IWS affects human health and wellbeing, as water is stored for long periods of time and water quality issues arise (Galaitsi *et al.* 2016).

It is important to recognize the problems of intermittent supply and to make an effort to find realistic solutions. The vast majority of IWS systems could run 24×7 IF physical losses were reduced to low levels (because then it would not make much difference whether you operate 12 or 24 hours a day). Of course there are places where absolute water scarcity is so bad that there is simply no other way than extreme IWS.

However there are many disadvantages to IWS compared to continuous drinking water supply, such as:

3.2.1 Water contamination and health hazard

Intermittency entails a high risk of contamination, which creates substantial health hazards. The first route is the ingress of contamination through broken pipes or joints. Interruption of supply can create low pressures or even a vacuum condition in pipelines that last for a significant period of time. Consequently, the contamination readily enters through leak points on pipes. The situation is serious in cities with no or poor sanitation facilities for collection and disposal where sewage flows in open ditches close to distribution pipes. The bacteriological quality of an intermittent water supply is substantially lower than of a continuous service since it is difficult to keep proper chlorination level in the network. Potentially contaminated water, such as rainwater, sewage spills, latrine drainage, etc. may enter the system through the fractures in the pipes when supply is off and create severe health hazard conditions for which chlorination levels acceptable to consumers will not be sufficient to deal with such situations. Furthermore, the contamination at household tanks is another type of risk as such storage is rarely constructed under supervision of the water company or according to certain standards with prescribed prevention of contamination. According to a water quality test in Istanbul, Turkey, 24% of samples from consumer storage tanks were found to be positive for coliforms, while only 4% taken from the pipe network presented positive (Yepes *et al.* 2001). Contaminated water requires consumer treatment or the use of bottled water, higher doses of chlorine are needed, and oversizing of networks is needed to supply the necessary quantities in a shorter time (McIntosh, 2003). The effects of IWS on health are documented also in the South African case study presented in Section 10.5.1.

It is difficult to keep proper chlorination level in the network since there are no constant hydraulic conditions with the repeated emptying and charging of the network. In order to deal with such situation it is normal to significantly increase chlorination, which of course entails other dangers such as the potential creation of Trihalomethanes (THMs). THMs are formed as a by-product predominantly when chlorine is used to disinfect water for drinking and result from the reaction of chlorine with organic matter present in the water being treated. The THMs produced have been associated through epidemiological studies with some adverse health effects and therefore limits are set on the amount permissible in drinking water.

3.2.2 Water wastage

Consumers exposed to IWS conditions are likely to keep their taps open to obtain as much water as possible whenever the service resumes. In addition consumers usually remove the control valves that are installed in the ground and/or roof tanks in order to remove any flow restriction hoping to get larger volumes of water in a shorter period of time. Under these circumstances, consumers experiencing IWS are likely to waste more water than those who receive a 24 × 7 supply from the fear of not having sufficient water they will tend to store as much as possible, which is usually replaced by the fresh supply of the next day. Unfortunately, for the less fortunate consumers who do not have the means of installing ground and/ or elevated tanks are forced to manage with the small quantities that they have managed to store in their individual containers.

3.2.3 Systems do not operate as designed

Although systems are designed to operate 24 × 7, they are forced to operate intermittently with different flow conditions. There is frequent wear and tear on valves, so more manpower is needed for network repairs. Very low tariffs add to problems, since utilities that lack funds struggle to meet O&M costs (McIntosh, 2003). The quantity of water to be made available over 24 hours has to be made available in fewer hours in an intermittent system, which requires distribution pipes with larger diameters. According to Yepes *et al.* (2001), once intermittent service becomes the norm, the hours of service continue to decline. The high costs of intermittent supply are paid by the utility, which incurs higher investment and operating costs. Pumping costs are reduced under an IWS scheme, but sub-optimal pumping leads to reservoir capacities being underutilized. What is probably not widely recognized is that pumping times are drastically reduced when balance storage is constructed and metering, billing, and collection controls are set, under a 24 × 7 operating scheme.

3.2.4 Inequitable distribution within the network

Another issue that is raised with the use of IWS is the inequitable distribution within a network. In distribution systems designed to operate for 24-hour supply, consumers exposed to intermittent supply conditions are likely to keep their taps open to obtain as much water as possible whenever the service resumes (IWWA, 2000). As the amount of water flowing out of taps comes to depend on the pressure head once the supply has been restored, peak flows that are larger than expected will occur in the pipelines, increasing the pressure losses in the network. Consequently, those consumers furthest away from supply points will always collect less water than those nearer to the source (Gottipati & Nanduri, 2014).

3.2.5 Inconvenience and high coping costs to consumers

Inconvenient supply times mostly affect the poor, since consumers have to pay for storage and pumping. Alternatively, they will have to go to public taps, sometimes quite faraway and even during midnight, to collect water. Long distances and queues are typical problems, as stated in Totsuka *et al.* (2004), of women and children from underprivileged areas, taking large amounts of productive time from them to wait for the supply to be turned on and to fill their small containers in order to secure a small quantity of water for their basic house needs as shown in Figure 3.2.

Figure 3.2 Consumers waiting to fill up their water containers (*Source*: Anand Jalakam, Bangalore, India).

As a consequence of intermittent supply, consumers have to pay the costs, so called *coping costs*, for additional facilities, such as storage tanks, pumps, alternative water supplies and household treatment facilities. The poor who cannot afford such facilities spend their time to fetch water from public taps or vendors at comparatively high total costs. The wealthy cope by spending money on water tanks, pumping systems and filters whereas middle-income groups spend less on capital equipment but more in terms of time and power. For the low-income group however the coping cost is primarily the opportunity cost of the time they must spend collecting water.

Ironically, such water often risks becoming contaminated and the cost in time often exceeds what wealthier consumers pay for water from house connections. Coping costs provide a lower bound to willingness-to-pay for water but are rarely quantified. Even those that argue that water should be provided free from public taps do not realize the significant economic costs in terms of time spent in carrying water. Choe *et al.* (1996) estimated the annual household coping costs through a sampling survey in Dehra Dun, India, and clearly showed that the coping costs paid especially by public tap users, who are usually the poor, are significantly high. Another study in Tegucigalpa, Honduras estimates that coping costs for poorest families with an intermittent household connection is around 180% of water tariff (Yepes *et al.* 2001). Consumers with intermittent supply may also spend money to obtain water of potable quality by boiling it or purchasing bottled water.

3.2.6 Seasonal dependence

Seasonality plays a significant role in IWS availability (McIntosh, 2003). In many regions, during the dry season, the availability of groundwater reduces two to three months before the rainy season begins, thus severely limiting water supply in already stressed areas. Seasonal populations also stress the water supply, as do electricity shortages, operations budgets, declining well yields, and adherence to fixed pumping schedules (Choe *et al.* 1996). The seasonality implication is expected to become more extreme under the pressure of climate change. Furthermore, if seasonal changes start to lead to IWS, then the network will deteriorate at a faster pace and a decade later, if no action is taken against network deterioration, water utilities will have to implement IWS all year long.

3.2.7 Network deterioration

Overall, IWS has a detrimental effect on the network, results in ineffective supply and demand management, inefficient operations, increased difficulties in detecting and fixing leaks as well as greater number of illegal connections.

IWS would cause inaccuracies in meter registration. Meter registers might reverse due to vacuum conditions created during emptying of the network as supply is cut-off. Air expelled from the pipes during filling might drive meters at excessive speed during the charging stage after the supply has been resumed resulting in the accelerated wear and tear of the registration mechanism. Undesirable environment, such as repeated dry and wet conditions, would accelerate the performance deterioration of water meters. Meter malfunction brings difficulties for water providers to monitor the water use and collect accurate tariffs. Furthermore, it makes consumers sceptical to the accuracy of their water bills relating to the meter registration (IWWA 2000).

3.2.8 Increased difficulties in detecting and fixing leaks

There are many difficulties in conducting leakage detection activities in networks under IWS and there are many parameters that ought to be considered prior to starting any leakage activities. Duration that the system is pressurised, pressure, pipe material are only some of the parameters that need to be considered prior to attempting any leakage activities.

It is imperative that conducting leakage activities in intermittent supplies using acoustic methods that the system must have some pressure and the higher the better. For mains that are buried at least 1 metre deep a pressure of at least 10 metres is desirable. The water supply has to be in the system long enough to fill all customers tanks and each tank has to be controlled via a ball valve or other mechanism so that when full there is no flow through the customer service connection otherwise it would interfere with the acoustic process indication showing that each property has a potential leak on it and would have to be investigated hence causing Active Leakage Control to be long and non-productive process.

In Chapter 7 a detailed account of leakage detection under intermittent water supply conditions is provided.

3.2.9 Meter malfunctioning & accelerated wear & tear

IWS would cause inaccuracies in meter registration. Meter registers might reverse due to vacuum conditions created during emptying of the network as supply is cut-off. Air expelled from the pipes during filling might drive meters at excessive speed during the charging stage after the supply has been resumed resulting in the accelerated wear and tear of the registration mechanism. Undesirable environment, such as repeated dry and wet conditions, would accelerate the performance deterioration of water meters. Meter malfunction brings difficulties for water providers to monitor the water use and collect accurate tariffs. Furthermore, it makes consumers sceptical to the accuracy of their water bills relating to the meter registration.

On the whole, IWS has a detrimental effect on the network, results in ineffective supply and demand management, inefficient operations, increased difficulties in detecting and fixing leaks as well as greater number of illegal connections. A detailed analysis of the problems associated with meter measurements under IWS conditions is given in Chapter 8.

3.2.10 Inefficient operations – more manpower

Numerous complaints are usually received from disgruntled consumers regarding quality problems and of course lack of pressure during intermittent supply. Needless to say that intermittent supply causes serious disruption and upheaval to the daily activities of people whether they at home or at work.

Intermittent supply has direct financial cost to the utility. On the one hand they suffer loss of revenue due to the decrease in water sales and on the other hand incur additional expenses paid to staff in overtime for opening and closing valves to implement water rationing and to repair breaks caused to the pipes due to the frequent emptying and filling of the pipes.

It is evident that more personnel is needed to run a system under IWS having to respond to emergencies and extreme operational conditions.

3.3 REFERENCES

Biswas A. K. (1992). Sustainable water development: a global perspective. *Water International*, **17**, 68–80.

Collier P. (2012). Bottom Billion. The Wiley-Blackwell Encyclopedia of Globalization. John Wiley & Sons, Ltd.

Collier P. and Venables A. J. (2012). Land deals in Africa: pioneers and speculators. *Journal of Globalization and Development*, **3**(1), 3.

Gardner-Outlaw T. and Engelman R. (1997). Sustaining Water, Easing Scarcity: A Second Update. Population Action International, Washington, DC, USA.

Gleick P. H. (1996). Basic water requirements for human activities: meeting basic needs. *Water International*, **21**(2), 83–92.

Gottipati P. V. and Nanduri U. V. (2014). Equity in water supply in intermittent water distribution networks. *Water and Environment Journal*, **28**(4), 509–515.

Hinrichsen D., Salem R. and Blackburn R. (2002). Meeting the Urban Challenge. Population Reports, Series M, No. 16. The Johns Hopkins Bloomberg School of Public Health, Population Information Program, Baltimore.

Indian Water Works Association (IWWA) (2000). International Seminar on Intermittent Drinking Water Supply System Management: Technical Papers. In: Proceeding for the International Seminar on Intermittent Drinking Water Supply System Management, Mumbai, India.

Jacobsen M., Webster M. and Vairavamoorthy K. (2013). The Future of Water in African Cities. Why Waste Water? International Bank for Reconstruction and Development, The World Bank. DOI: 10.1596/978-0-8213-9721-3.

Khroda G. (1996). Strain, social and environmental consequences, and water management in the most stressed water systems in Africa. In: Water Management in Africa and the Middle East-Challenges and Opportunities, IDRC, Canada, p. 295.

Maplecroft (2010). Climate Change Vulnerability Index. http://www.maplecroft.com.

McIntosh A. C. (2003). Asian Water Supplies. Reaching the Urban Poor: A Guide and Sourcebook on Urban Water Supplies in Asia for Governments, Utilities, Consultants, Development Agencies, and Nongovernment Organizations. Asian Development Bank and International Water Association.

Rijsberman F. R. (2006). Water scarcity: fact or fiction? *Agricultural Water Management*, **80**(1), 5–22.

Rosegrant M. W., Davila-Poblete S., Dawe D., Elliot H., Kaosa-ard M., Meinzen-Dick R., Palanisami K., Pingali P., Samad M., Swallow B. and Wolf A. (2002). Policies and institutions for sustainable water resource management: a research agenda. Challenge Program on Water and Food background paper 5.

Serageldin I. (1995). Toward Sustainable Management of Water Resources. The International Bank for Reconstruction and Development, The World Bank, Washington, DC.

Shiklomanov I. A. (1991). The world's water resources. In: Proceedings of the International Symposium to Commemorate 25 Years of the IHP, UNESCO/IHP, Paris, France, pp. 93–126.

Singh V. P. (2000). The entropy theory as a tool for modelling and decision-making in environmental and water resources. *WATER SA-PRETORIA*, **26**(1), 1–12.

Totsuka N., Trifunovic N. and Vairavamoorthy K. (2004). Intermittent urban water supply under water starving situations. In: Manuscript Prepared for the 30th WEDC International Conference on People Centered Approaches to Water and Environmental Sanitation.

UNESCO (2003). Water for People Water for Life – The United Nations World Development Report. United Nations Educational, Scientific and Cultural Organisation, New York, USA.

WHO (2003). Water Sanitation and Health. http://www.who.int/water_sanitation_health/hygiene/en/

WHO/UNICEF Joint Water Supply, Sanitation Monitoring Programme, World Health Organization, & UNICEF (2008). Progress on Drinking Water and Sanitation: Special Focus on Sanitation. World Health Organization.

Yepes G., Ringskog K. and Sarkar S. (2001). The high cost of intermittent water supplies. *Journal of Indian Water Works Association*, **33**(2), 167–170.

Chapter 4

Reasons for resistance to change towards 24 × 7

Bambos Charalambous and Patricia Bakir

There is much confusion to the public about the pros and the cons of intermittent supply, probably due to lack of data. Even though it is not true, there is the belief that for older distribution systems with weaker joints and more leakage, restrained supply hours can limit leakage (Sustainable sanitation and Water Management). The thought is that IWS leads to lower pressures, thus helping to lower network leakage. On the contrary, according to (McIntosh & Yñiguez, 1997; De Marchis *et al.* 2011 and Fontanazza *et al.* 2007, 2008), systems do not operate as designed with intermittent supply, which leads to network deterioration and meter malfunctioning (IWWA 2000). According to Fontanazza *et al.* (2007), intermittent distribution has the advantage of requiring small financial efforts, but with such a statement, the high coping costs of IWS are overlooked (Totsuka *et al.* 2004). The availability of time for repair and maintenance of the network is listed as a benefit, since supply hours are limited; also, a prevailing thought is that scarcity is actually dealt with by interrupting the water supply and thus saving resources.

Despite low levels of the IWS service, the arguments against the idea of continuous 24 × 7 water have been strong: "*We don't have enough water in a water-scarce country to supply continuous water*"; "*It will cost too much when so many people are poor and tariffs are already too low*"; "*We have intermittent power supply so how can we expect to have continuous water supply*?"; and "*Our cities are growing too quickly to support continuous water*". Some of these questions seem reasonable, while others stem from lack of information or misinformation that dominates especially in the poor regions.

With a few exceptions, users of piped water distribution systems in Indian cities receive an intermittent supply of water. Discontinuous supply is practiced even though many cities have sufficient water resources to provide a continuously pressurized system, operated 24 hours a day. Although the Indian Manual of Water Supply and Treatment (1999) recommends that intermittent supply be discouraged, its practice has

Bambos Charalambous and Chrysi Laspidou
doi: 10.2166/9781780407074_029

become the standard; thus, it is now considered the norm in India, rather than a sub-standard exception. Some factors that have contributed to this practice of discontinuous supply include (WSP, 2003).

- Rapid growth in population and water demand (and in some areas, shortage of water);
- Inadequate water charges and billing/collection mechanisms, leading to insufficient revenues to repair, maintain, and replace infrastructure;
- Intermittent and poor quality electricity supply;
- Inadequate human resource development, including training in modern utility operations; and
- Inadequate demand-responsiveness and customer-orientation among service providers (WSP 2003).

One of the strongest arguments is that the immediate cost-benefit of moving to continuous supply is not as clear-cut once much of the capital component of users' coping costs (rooftop tanks, pumps etc.) has been sunk. This may help explain why existing customers, esp. middle-class, are not more vociferous in demanding reform.

Given strong arguments against 24 × 7 supply, it is clear that developing a successful demonstration project will need not only technical skills but also significant communication and social skills. Strong leadership is an important requirement, especially at the government, political, and official levels; the support of decision makers in financing agencies, academic institutions, and engineering departments is also essential.

4.1 THE UTILITY POINT OF VIEW

Intermittent water supply is practiced by many water utilities round the globe and in many cases it has been the normal modus operandi for many years, in fact it has become the norm and there is strong resistance in many cases in even considering moving to 24 × 7.

A common reason for resistance is the fact that utilities are used to "business as usual" and do not wish to change this perhaps through fear of the unknown. If utility staff do not understand and appreciate the need for change it is expected that they would strongly resist particularly those who feel that there is nothing wrong with IWS and the current way of doing things; it works well… and has done for all these years, why change it!

People in the utility have been doing things for many years in a certain way and if they are asked to do these in a new way, as rational as that new way may seem, they will resist perhaps due to fear of the unknown which they will seldom admit. Change from IWS to 24 × 7 will necessitate changes in skills, knowhow, competencies and some people will feel that they will not be able to make the transition very well and therefore resistance to change is the obvious path for them.

However change would not only require staff that is willing to change and acquire new skills and competencies but would also need to be given the means both technical and financial in order to achieve the change to 24 × 7. Without CAPEX to build DMAs, replace dilapidated mains and service connections, etc. and of course OPEX to cover the substantial increase in the cost of leak detection and repair any efforts to move from IWS to 24 × 7 will be opposed primarily by the staff who will rightfully feel that such a crucial change could not be successfully be made.

In general utilities are cumbersome, conservative organisations, which resist change in the status quo and resistance can stem from perceptions of the change that people hold. People feel that they could be worse off at the end of the change and therefore because they may have to be doing more work or have more responsibilities.

Expecting resistance to change and planning proactively for it from the beginning of the change process will allow an efficient and effective transition from IWS to 24 × 7.

4.2 THE CUSTOMER POINT OF VIEW

"Even though the change is meant to be beneficial, if customers are accustomed to receiving a service a certain way, any change has to be carefully managed."

This sentence from the Sri Lanka Case Study (see Section 10.6) sums up the challenge of customer acceptance to moving to 24 × 7 supply. If customers are receiving water a few days – or even a few hours–a week, one would expect them to be fully supportive of any program to move from IWS to 24 × 7 supply. They will have water available all the time for all their needs, not need any storage facilities in their homes and have better water quality due to stable hydraulic conditions in the network thus avoiding ingress of contaminants into the pipeline when empty. However, there are some misconceptions and some hidden consequences, which need to be taken into account by water service providers.

In areas where there is intermittent supply, customers have coping mechanisms: storage tanks, buying from private vendors, household water conservation measures, on-site boreholes (which may also be used to sell water), and yes, stealing water from the municipal supply network. Keeping in mind that customers need to be segmented into household, industry and businesses, and even sub-divided further – this section outlines the main reasons customers may support and may be against adoption of 24 × 7 supply. Chapter 9 will address how to communicate with them and customer segmentation.

Transitioning to 24 × 7 supply will almost always include a major NRW program – which requires reduction in both physical and commercial losses. What does that mean for customers? It means major construction projects in their neighborhoods, which can be very disruptive on a short-term – or even long-term basis. It also can mean supply stoppages during infrastructure upgrading. It also means a push to increase bill collection and reduce water theft – something not all customers will be happy about. Here is a list of questions that shows how customers perceive a transition to a 24 × 7 operational regime.

- Will water rates go up to pay for the cost of transitioning to 24 × 7? Will they go up even before 24 × 7 is in place? Customers will find it very hard to support this unless they trust the water provider that their increased payments will provide better and sustainable service.
- Who gets 24 × 7 first? Will it be step by step, one area after the other? If not managed correctly – both technically and with good communication – customers in areas not receiving 24 × 7 may get angry that they have not yet received it. Tariff structures would normally need to be reviewed and increase accordingly on an area by area basis as 24 × 7 is introduced to each area which of course needs to be communicated to the customers from the very beginning making them aware that the improved service that they will receive will have an additional cost.
- Will water quality be better or worse? Water quality with 24 × 7 supply should be better, but there may be an initial period of increased leaks and pipe bursts possibly causing quality and supply problems. There may be some discoloration of water due to leakage repairs, which will upset customers. This can be alleviated easily through a good communication program.
- Does 24 × 7 also mean direct to all households? Does that mean communal standpipes will be discontinued? This means individual households will have to connect to the network – a cost to them they may not be able to bear. This is where the service provider should consider bearing the cost of such connections.

Customers may use more water when they go from having it irregularly to 24 × 7. This can be just for a short period of time until they adjust to the new supply, but can cause concern due to higher bills, and also cause problems for the utility in supplying enough water. Their bills may go up even more than expected in cases where there is a block tariff and they jump to a higher use tier. Other customers may be used to obtaining water illegally - some even to sell it to others. They will resist any changes, which make that unnecessary.

In many instances, poorer customers do not receive enough water to meet minimum health needs due to high cost and/or supply shortages and IWS. So an increase by some customers in water use is in fact desirable to improve public health. Also water conservation programs directed at all customers can be offensive to those who have always had very little water. This is just one example of how it is extremely important to segment customers when planning to move to a 24 × 7 supply. Some customers will actually want to use more water, but not too much more due to costs and awareness campaigns to help them continue to use the water wisely will be needed. Programs for household water conservation at a wider level will need to be carefully targeted to higher consumption customers who may already be wasting water and do not have cost constraints. These customers are the ones most likely to use much more water when there is a 24 × 7 supply.

What if 24 × 7 is promised and provided for a time, then revert back to IWS? This is a huge problem, as documented in the Ghana pointed out in the Ghana case study (see Section 10.3). In this case, the government promised 24 × 7 supply, and started to provide it, but then the contractor reverted to intermittent supply. Raising customer expectations is never a good thing for any business. In addition, customers may have removed their water storage tanks in their household; and new households will not have storage systems at all. This will greatly reduce trust in the water service provider and cause undue harm to customers.

Upgrading and replacing water meters is also important. If the existing customer water meters are not functioning properly and 24 × 7 supply is introduced there will be billing errors, resulting in lower revenues for the water utility.

Most of these issues can be alleviated through strong communication with customers. Politicians are always averse to raising prices and often will provide greater resistance to moving to 24 × 7 than customers because they do not want to be unpopular. Many studies worldwide have shown that customers are willing to pay for services – if they get the promised service.

There also may be vested interests in keeping IWS. Sometimes private vendors actually tap into public water supply systems, stealing water, which they then sell to consumers who do not receive enough water. There are also of course perfectly legal water vendors who also may lose business with a 24 × 7 system– some of these may buy water from the municipal supply when it comes, then sell on off days. In many countries, private vendors are a very strong lobby and can resist in both legal and illegal manner. They also are often residents of a local area, so have the ability to stir up customers against the new system.

Will prices go up? If they are going to go up, many studies have shown that customers are willing to pay – but only if they are really sure they are getting better supply.

Usually, moving from IWS to 24 × 7 is part of a comprehensive program to reduce NRW, including commercial losses due to lack of billing or meter under-registration. If customers have been receiving IWS but been getting away without paying bills, they may resist having to pay bills.

4.3 OTHER STAKEHOLDERS AND VESTED INTERESTS

In countries where IWS is a long established situation the water supply can be classified into two categories – formal and informal. A formal system normally refers to delivery through a utility's piped system. An informal system of water supply usually includes alternative water delivery mechanisms and practices, which are largely unregulated by the utility and/or the local government or the state.

It is normal that if a utility fails to deliver the level of service that the people require, that is good quality water in reasonable quantities, the users will seek alternative sources. This need creates a demand, which is quickly met by private initiatives, which in many cases expand to considerable size businesses.

The informal water suppliers are driven by the desire of the users to secure satisfactory quantities of water for their needs and to some extent for their convenience. Financial investments are made by the informal water suppliers in order to be able to respond to the needs of the users. Therefore moving from IWS to 24 × 7 will adversely affect profitable businesses that have been set up to fill a gap, which the water utility will move back to fill and this may cause serious resistance by the affected parties.

4.4 REFERENCES

De Marchis M., Fontanazza C. M., Freni G., La Loggia G., Napoli E. and Notaro V. (2011). Analysis of the impact of intermittent distribution by modelling the network-filling process. *Journal of Hydroinformatics*, **13**(3), 358–373.

Fontanazza C. M., Freni G. and La Loggia G. (2007). Analysis of intermittent supply systems in water scarcity conditions and evaluation of the resource distribution equity indices. *WIT Transactions on Ecology and the Environment*, **103**, 635–644.

Fontanazza C. M., Freni G., La Loggia G. and Notaro V. (2008). Definition of Performance Indicators for Urban Water Distribution Systems in Drought Conditions. IWA Publishing, London, UK, pp. 35–46.

Manual of Water Supply and Treatment (1999). Ministry of Urban Development, Government of India, New Delhi, 3rd edn, p. 359.

Sustainable Sanitation and Water Management. http://www.sswm.info

Totsuka N., Trifunovic N. and Vairavamoorthy K. (2004). Intermittent urban water supply under water starving situations. In Manuscript Prepared for the 30th WEDC International Conference on People Centered Approaches to Water and Environmental Sanitation.

WSP (2003). 24-hour Water Supply: Is this Goal Achievable? Drawing lessons from rapid distribution system diagnostic assessments in Indian cities. In: Fifteenth Meeting of the Urban Think Tank, 23–24 September 2003, Nagari, Hyderabad, India.

Chapter 5

Understanding water losses in the context of IWS

Bambos Charalambous and Roland Liemberger

5.1 THE CHALLENGES IN HAVING IWS

In many world regions, Intermittent Water Supply (IWS) systems are prevalent. Although intermittent water supply may seem to be a solution to a water shortage situation, in overall terms the water balance is adversely affected. Supplying less quantity in an intermittent manner causes such deterioration to the network that when continuous supply is re-established additional quantities are lost through increased leakage, which in fact places an added financial burden on the water utility. It is therefore evident that no matter how good a network is, intermittent supply operation definitely has a detrimental effect on its integrity; in addition, the amount of water "saved" is later "lost" and in greater quantities through increased levels of leakage. Such operational conditions should be avoided especially in pipeline networks that have been designed for continuous supply.

It is worth noting that IWS is enforced not only in cases where there is water shortage but also where the hydraulic capacity of distribution networks is such that cannot satisfy demand as well as in cases where the network is severely deteriorated resulting in high leakage.

It has been shown that domestic demand is in effect inelastic, if the water utility is to maintain a reasonable level of service to the customers in terms of water quantities provided during an IWS regime. In fact, water saved by customers under such operating regimes is on the whole very little. It is generally acceptable that better results could be achieved through a structured conservation programme rather than intermittent supply. Of course, such programmes need to be introduced as part of an overall strategy for water conservation both on the supply and demand side

In many instances there is no indication how long intermittent supply will be in place. In many countries around the world, IWS is the norm rather than the exception e.g. India (see box). The hydrological

Bambos Charalambous and Chrysi Laspidou
doi: 10.2166/9781780407074_035

conditions in each case could adversely impact water supplies for years, in which case conserving as much as possible the limited water resources may not be the long term solutions but it may be necessary to add to the water balance new non-conventional water resources. In many countries, water shortage problems were overcome through the desalination of brackish or saline water. Of course exploring every potential water source available may be the only solution in many instances, but leakage reduction is always one of the least expensive and quickest solutions to ensure that water will be available when needed.

It is generally considered that IWS is not an ideal form of supply and should not constitute a permanent solution; however, it is applied by many water utilities with great ease mainly as a measure for dealing with water shortage or drought conditions without seriously looking into alternative solutions. It is also the authors' experience that some water utilities are applying IWS as a measure to reduce extremely high leakage from their networks, which of course prevents them from maintaining a continuously pressurised network with all the adverse repercussions.

Even though in a number of instances it may not be possible to avoid IWS, the advantages of IWS if any, are very few and lack substance in order to convince that the use of intermittent supply is a sustainable *modus operandi* for water distribution networks. Intermittent supply is usually introduced either as an emergency measure, when the water availability is limited or in some cases it is introduced as a measure to control water use and to reduce leakage. In the first case, when there is limited water availability, there may be no alternative to the rationing of water and an intermittent supply cannot be avoided once the supply resource has been depleted. In the second case, however, where the intermittent supply is introduced as a water saving measure, there may well be alternative interventions that can provide savings without some of the problems that tend to accompany such pressurising and depressurising of the distribution network (Mckenzie, 2016).

In many systems, IWS was not an element of initial system design but rather reflects a combination of deteriorating infrastructure and demand growing beyond design limits. A possible combination of factors such as: water scarcity, prolonged drought periods, population growth, urbanisation and increasing demand, lack of awareness and forward planning may have been the root causes of IWS for many water utilities (see Chapter 3). Inevitably, IWS is the cause of serious problems in the proper operation and management of a water distribution network.

IWS is a major issue in India and is closely linked to water loss from the distribution networks. The main challenges in managing water loss in systems with IWS include pressure fluctuations that compromise the structural integrity of the network, an increase in pipe failure and network deterioration, increased difficulties in detecting and fixing leaks, water contamination and public health issues, illegal connections and customer meter malfunctioning to name a few.

More than a third of India's urban population has no access to safe drinking water, and the majority of urban residents struggle to cope with intermittent water supply conditions. The average number of hours per day (24 hours) across a number of urban centres in India is ranging from as low as 1 hour per day to 18 hours per day with a weighted mean of 4.5 hours per day.

5.2 THE VICIOUS CYCLE OF IWS

Normally water reticulation networks are designed to provide piped water on a continuous basis without any discontinuity in this supply apart from extremely short intervals where the supply is cut-off for routine maintenance or fixing of pipe breaks. However, in some instances changing hydrological conditions may result in water shortage and the water utilities are unable to meet existing needs. In some geographical areas this situation may take the form of a cyclic phenomenon where periods of low rainfall are repeated every few years, resulting in water rationing and the temporary application of IWS (Charalambous, 2009) applied as a measure to deal with such circumstances.

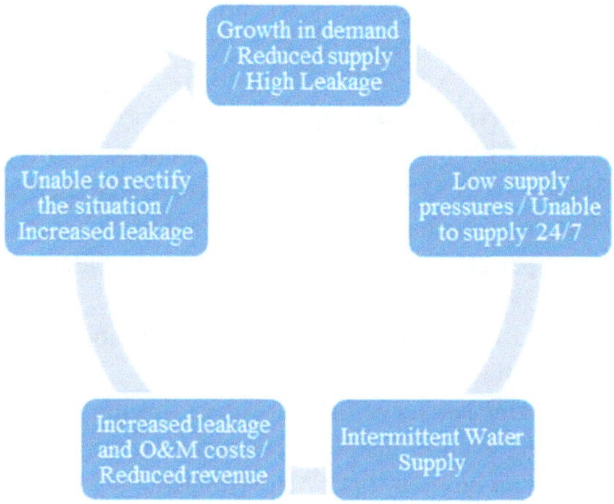

Figure 5.1 Vicious Cycle of IWS.

5.3 MYTH BUSTERS

Over the years, a number of misconceptions have been linked to IWS, particularly relating to leakage and customer consumption. Based on their own experiences as well as data and information provided by colleagues, the authors' set out below evidence which clearly shows that the "myths" build around IWS are just not true, such as under an IWS regime the NRW is lower compared to 24 × 7 supply or the volume of water distributed under IWS is less compared to 24 × 7 regime. Analytically the "myth busters" are presented below:

Is distributed water less under an IWS regime? It has been considered that under IWS conditions the volume of distributed water is less than the volume needed under a 24 × 7 supply regime. However, evidence from the Karnataka Demonstration Project (Jalakam, 2014) demonstrated that this is not the case as it can be seen from Figure 5.2.

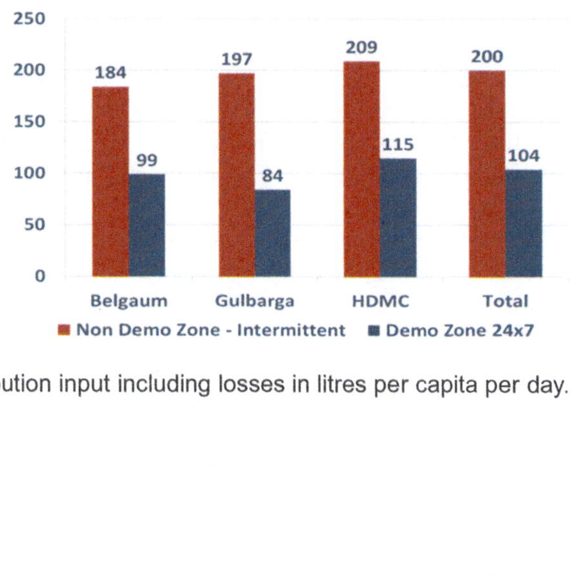

Figure 5.2 Water distribution input including losses in litres per capita per day. (*Source*: Anand Jalakam, Bangalore, India).

The volumes of water that were distributed to the demonstration zones in each city were far less compared to the volumes that were distributed to the areas of each city under IWS. In fact the numbers show that on average for the 3 cities the volume distributed to the network under IWS was the equivalent of 200 litres per capita per day compared to 104 litres per capita per day for the demonstration zones which were under 24 × 7 regime, that is 50% less water on average was distributed under the 24 × 7 regime in the demonstration zones.

Is IWS an effective leakage reduction measure? Data and information relating to leakage have been collected and analysed for a distribution network, which was operated over a 2-year period under an IWS regime (Cyprus case study, Section 10.2). Detailed analysis showed that there was a significant increase in the Minimum Night Flow, which was attributed to the additional breaks that the network suffered during the 2 years of intermittent supply period, proving that IWS is not an effective leakage reduction measure; on the contrary, it increases pipe breaks and overall network leakage. Further analysis showed that the number of breaks on mains increased by 200%. Similarly the number of reported service connection breaks increased by approximately 100%.

Is IWS an effective drought/water conservation measure? Taking into account the fact that an IWS regime results in increased leakage, even if customer consumption remains the same (it is often seen to increase due to water wastage), it can be seen that IWS will ultimately result in using increased volumes of water for water supply, when compared to 24 × 7 operation. Thus, clearly, IWS is not an effective water conservation measure – in many cases, it will result in the opposite effect.

5.4 THE CHALLENGE

Water Service Providers in many countries around the world that run their networks on intermittent supply are in need of assistance to improve their services. The key to better service is generally enhanced and improved management of systems, not just more capital expenditure on assets, yet for the vast majority of these Water Service Providers there is no clear route to management reform. Traditional training and capacity building approaches are indeed necessary to build a solid foundation on which the Water Service Providers will achieve long term sustainability and improved level of service, while Public Private Partnerships shall provide the vehicle for transitioning from IWS to 24 × 7 in an efficient and effective manner. However, in many countries this approach may still not be feasible or politically acceptable but it has generally being gaining wide acceptability in recent years due to successful applications.

While it is relatively easy to turn a 24 × 7 system to an intermittent supply, it is very hard to do the opposite. Water utilities that have fallen into the vicious cycle of IWS have major institutional, technical and financial issues and would definitely need to go through a reform process; moving to continuous supply requires often very difficult political and institutional choices that many water utilities/governments prove reluctant to make. A paradigm shift is therefore imperative to transition from IWS to 24 × 7 supply.

In order to improve operational, commercial and institutional efficiency the water utilities will need to strive towards reducing their water losses in the first instance coupled with an increase in the hours and days of supply until continuous supply conditions are achieved. A final step in this process, once low water loss levels with continuous supply are achieved, is to reduce and sustain the level of water losses to an economic level.

5.5 THE NEED FOR A STANDARDISED APPROACH

Before the first edition the IWA manual of Performance Indicators (Alegre *et al.* 2000) was published, there was no international attempt to standardize the water balance and water loss performance indicators. The IWA water balance (Liemberger, 2005) and water loss PIs have meanwhile become international standard and are promoted by many regional and national professional associations around the world (including AWWA).

In summary, the IWA water balance methodology and the IWA water loss PIs can also be used in IWS systems – IF the supply time is properly taken into account. Once the water loss situation is properly understood, forecasts can be made how much water will be required to supply the network in its present condition on a 24 × 7 basis and how much will be needed after network rehabilitation.

Details on the use of water loss PIs under IWS conditions and recommendations for transitioning to 24 × 7 are detailed in Chapter 6 of this book.

5.6 REFERENCES

Alegre H., Baptista J. M., Cabrera E. Jr., Cubillo F., Duarte P., Hirner W., Merkel W. and Parena R. (2000). Performance Indicators for Water Supply Services. IWA Manual of Best Practice, 1st edn. ISBN 1-900222-2722.

Charalambous B. (2009). Water Crisis – Bridging the Gap. In: Water Loss 2009 Conference Proceedings, 26–30 April, Cape Town, South Africa, pp. 241–246.

Jalakam A. (2014). Experiences of Intermittent Water Supply in India. In: Presentation at the IWA Water Loss 2014 conference, 30 April-2 April 2014, Vienna, Austria.

Liemberger R. (2005). The New German Water Loss Regulations in context with other International applications of the IWA Water Balance and real loss performance indicators. In: Paper presented at IWA Conference, Chile.

Mckenzie R. S. (2016). The Dangers of Intermittent Supply as A Measure to Save Water in South Africa. In: Paper to be presented at the 80th Conference of The Institute of Municipal Engineering of Southern Africa (IMESA), 26–28 October 2016, East London, South Africa.

Chapter 6

Transitioning from IWS to continuous 24 × 7

Bambos Charalambous, Jan Janssens and Roland Liemberger

6.1 APPROACHES TO WATER LOSS REDUCTION

Approaches towards water loss reduction most often tend not to look at the organization as a whole, but rather narrowly at the components of NRW in isolation (Charalambous, 2005). Frequently neglected in the design of the interventions, are the following key issues:

(1) the dynamic nature of NRW and the interactions of its components;
(2) the operator's capacity to cope with the intervention;
(3) the change process that needs to take place in the organization to internalize the intervention and secure the gains in performance in the long term.

Based on these pillars, reduction of water loss requires a structured holistic approach built on a long-term strategic plan which entails the following three phases:

- *"REDUCE"*: A first phase which comprises major interventions geared toward significant NRW reductions, breaking the vicious cycle with major investments in leakage reduction as appropriate.
- *"FINE TUNE"*: A second phase cantered in fine tuning the operational performance of the utility towards reaching the economic NRW level, and,
- *"SUSTAIN"*: The third and final phase to maintain the utility's performance at the economic level of NRW with minor, but annual recurrent investments with full cost recovery.

Starting from the concern of how making gains of water loss efficiency sustainable, recent experiences show a tendency to move away from short-term interventions or "business as usual".

6.2 ASSESSING WATER LOSSES

Control and efficiency of a water distribution network following an investment in the water resources and distribution system is imperative in today's world climate. Following the initial investment into providing

Bambos Charalambous and Chrysi Laspidou
doi: 10.2166/9781780407074_041

a potable water network, successful management of the infrastructure is required in order to operate as efficiently and cost effectively as possible. The management of Non-Revenue Water (NRW) levels is a key element to achieving this target. Although the assessment of NRW and water losses in general is covered in many other credible publications, a brief description of the tools developed by the Water Loss Specialist Group of the International Association is presented in this section for ease of reference and completeness.

Water Losses, Real (Physical) and Apparent (Commercial) occur in all water distribution systems – it is only the magnitude that varies. The level of NRW varies from country to country, region to region, utility to utility and will be affected by pipe network characteristics, metering, operational practice, level of technology and expertise applied to controlling water losses and other local factors. The components of water losses will also vary and for the NRW professional it becomes essential to understand the relative significance of each component before formulating a NRW strategy.

Significant contribution to reaching the point of water accountability was the establishment of the IWA Water Balance (Figure 6.1) that is a useful tool in analysing the various components of water production, storage and distribution. Through this analysis the magnitude of the water loss problem is identified and priorities are set for rectifying the situation based on the component analysis of the Revenue and Non-Revenue Water elements.

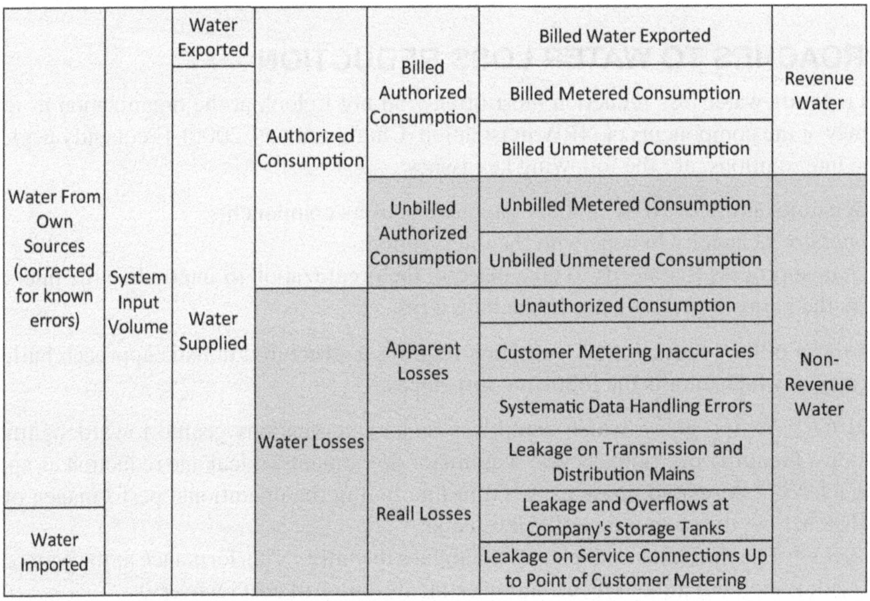

Figure 6.1 IWA Water Balance.

Abbreviated definitions of principal components of Figure 6.1 are as follows:

- *System Input Volume*: the annual input to a defined part of the water supply system.
- *Authorized Consumption*: the annual volume of metered and/or non-metered water taken by registered customers, the water supplier and others implicitly or explicitly authorized to do so. It includes water exported, and leaks and overflows after the point of customer metering.
- *Non-Revenue Water (NRW)*: the difference between System Input Volume and Billed Authorized Consumption.

- NRW consists of Unbilled Authorized Consumption and Water Losses.
- *Water Losses*: the difference between System Input Volume and Authorized Consumption, consisting of Apparent (Commercial) Losses and Real (Physical) Losses.
- Apparent (Commercial) Losses consists of Unauthorized Consumption and Customer Metering Inaccuracies.
- *Real Losses (Physical)*: the annual volumes lost through all types of leaks, bursts and overflows on mains, service reservoirs and service connections, up to the point of customer metering.

NRW comprises Unbilled Authorized Consumption, Apparent (Commercial) Losses and Real (Physical) Losses. These components are briefly described below:

- Unbilled Authorized Consumption is water used legally and to the full knowledge of the water utility but is not billed and therefore does not produce any revenue. Examples are water that is used for firefighting, street cleaning, drain flushing, municipal buildings and public gardens, etc. This component must be examined carefully by the water utility and minimized to the maximum possible degree.
- Apparent (Commercial) Losses have been clearly defined as influenced by four components, and that these components can act and interact interchangeably (Figure 6.2).

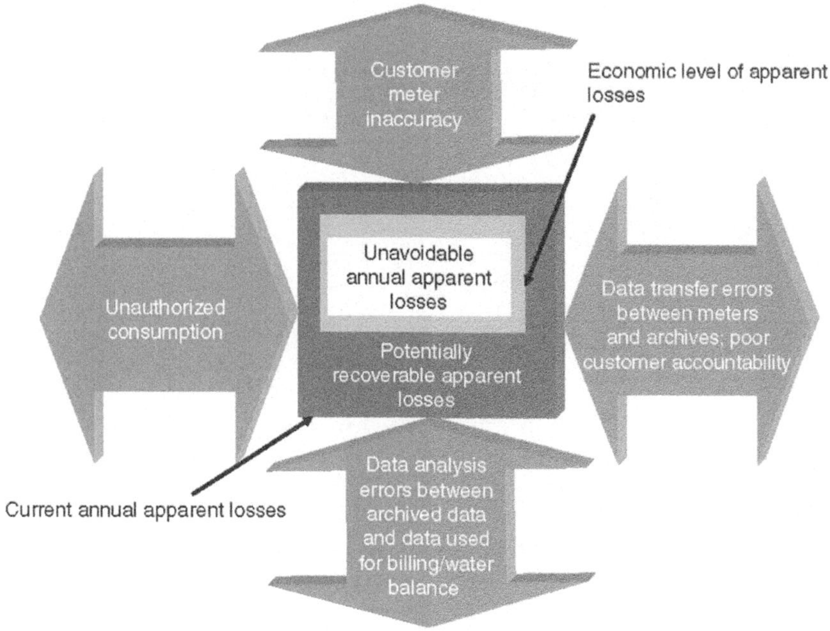

Figure 6.2 The Main Influences on Apparent (Commercial) Losses.

The main influences on Apparent (Commercial) Losses are:

- Meter under-registration is the inability of a revenue meter to accurately measure water, especially at low flows. This tends to increase with time and as the meter degenerates.
- Water theft is easy to conceptualize and consists usually of bypasses to the water meter, illegal connections, or willful damage to the water meter.

- Meter reading errors consist of genuine mistakes or intentionally incorrect meter reading.
- Water accounting errors consist of billing anomalies, such as computer-based estimations that do not reflect actual consumption values.

This approach will assist utilities in reducing their Commercial Loss component thus increasing utility revenues.

Real (Physical) Losses can be reduced by four leakage control strategies:

- Active leakage control
- Pressure management
- Speed and quality of repairs
- Targeted renewal of infrastructure

These four strategies need to be balanced in order to achieve the most cost effective leakage program that reduces leakage to an economically, environmentally and socially acceptable level (Figure 6.3).

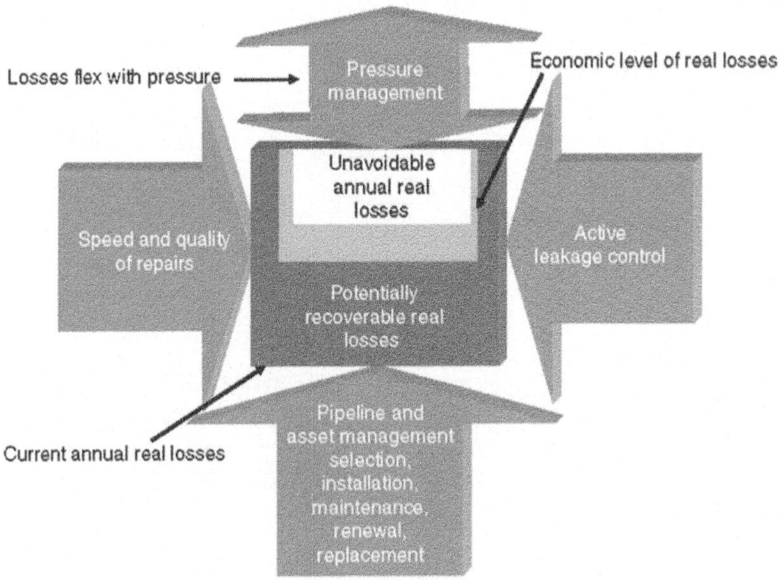

Figure 6.3 Four Real (Physical) Losses Control Strategies.

Physical Losses cannot be eliminated totally. The lowest technically achievable annual volume of Physical Losses for well-maintained and well-managed systems is known as Unavoidable Annual Real Losses (UARL). This leads to the definition[1] of the Infrastructure Leakage Index (ILI), which is a performance indicator for the benchmarking of physical losses.

[1]The Infrastructure Leakage Index (ILI) is the ratio between the current volume of real losses and the unavoidable volume of real losses at a specific supply time (h/d). An ILI of 1 means that the current real losses are as low as the unavoidable (minimum) volume of real losses. An ILI of 50 means that the current real losses are 50 times as high as they would be in a well-managed system in excellent condition.

6.3 MOVING AWAY FROM "BUSINESS AS USUAL"

The primary stage in designing a water loss reduction strategy is to quantify water uses and losses and understand the network operation and performance. There is in general no one standard approach and the strategy must be tailored to each individual water utility. The strategy advocated for this Study comprises the following elements:

- Specific, to take account of all relevant factors
- Holistic to address all components of NRW, Commercial, Physical and Unbilled Authorized Consumption
- Baseline and Target Setting
- Auditing and validation procedures, both internal and external, to sustain the gains of efficiency

Practice has proven that a passive control policy is insufficient to moderate water losses. Crisis management can result in damage to roads and housing, loss of supplies, difficulty in manpower planning, premium costs for repair and the loss of large volumes of water. It is true to say that within any distribution system there is a "Natural Rate of Rise" in water losses. This is primarily caused by on-going deterioration of the water meters and of the network, new leaks and bursts, and in doing "business as usual". Therefore, a passive approach will result in a continuous rise of the NRW, as shown diagrammatically in Figure 6.4, with detrimental effect on the level of service to the consumers. Gradually this deterioration may result in intermittent supply and ultimately to a complete failure by the water utility to continue to provide the required service.

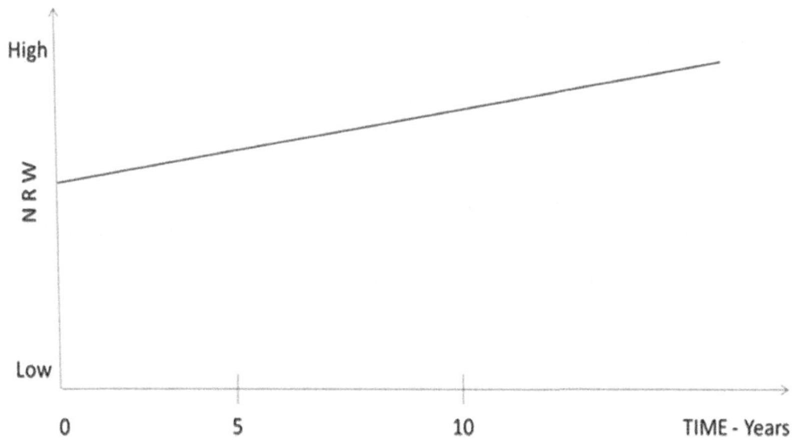

Figure 6.4 Passive approach to NRW.

To turn around this situation and to stop the vicious cycle of continuous increase of NRW levels, which ultimately will lead to intermittent water supply, it is imperative that a structured approach is established and that the required interventions are put in place in order to have the desired results for water utilities. This rationale is conceptualized and transformed into tangible and practical actions through structured interventions within the framework of a Non-Revenue Water long term strategic management plan as illustrated in Figure 6.5.

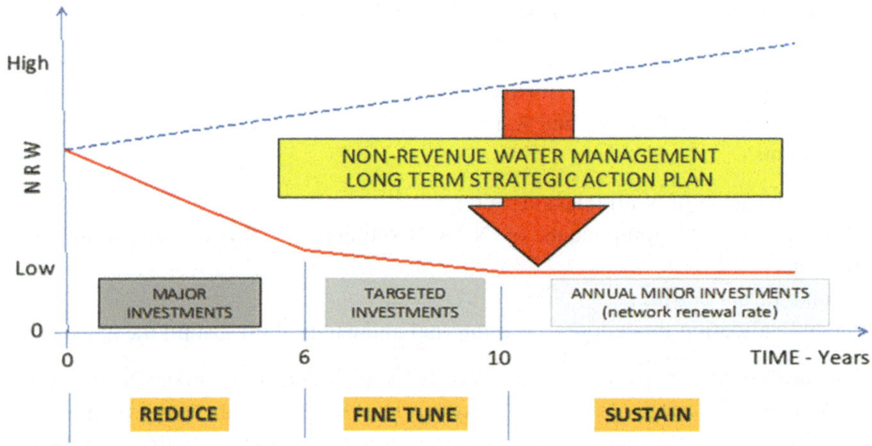

Figure 6.5 Structured Interventions in a Long-term NRW Strategy.

Reduction of NRW or specific components of NRW such as Real Losses or Apparent Losses or both requires a structured approach based on a long term strategic plan which will entail the following phases:

(1) The first phase will comprise major interventions, which will be geared towards reducing NRW from high to lower more manageable levels. This of course will require the corresponding investments in order to achieve the necessary results. A typical time period for achieving satisfactory results is 5–6 years depending on size, complexity and condition of the network.
(2) The second phase could be of shorter duration than the first phase, 4–5 years, and is centred in fine tuning the operational performance of the water utility by applying targeted investments in areas that will have the maximum impact in reducing NRW to an economic level as well as build up capacity within the water utility.
(3) The third and final phase is equally important as the first two. Having reduced the NRW to an economic level it is imperative that a framework is in place that will maintain this performance thus achieving sustainability of operations with only minor but annual recurrent investments.

Each of the above phases will have an immediate, medium or long-term strategy attached to it dependent upon their duration and whether they are one-off or on-going expenditures. The following time scales may be applied:

- Short Term or Immediate Measures – normally implemented in six months to one year
- Medium Term Measures – have time scales of one to five years
- Long Term Measures – have time scales of more than five years

For example, production metering will take less than one year to implement and is necessary for the assessment of the Annual Water Balance that will determine future investment strategies. Introduction or improvement of production metering is therefore seen as a "short term" measure, while maintain the quality, accuracy and reliability of production meters is an ongoing medium and long term measure.

It must also be recognized that individual options from the range of options to be applied may not necessarily be implemented in isolation. Additionally, the sequencing of options may be restricted by practical considerations. For instance, it is not possible to introduce effective pressure management into a distribution network where network zoning has not been implemented. These practical considerations impose a certain logical sequence to the implementation of the investment options. Whilst it is always desirable that the "least cost" options are carried out first, this cannot always be the case when the logical sequence imposed by practical considerations must be followed.

Successful change requires in addition to a structured innovative strategy a strong element of personal leadership. The challenge is how to "de-personalize" good utility performance and how to anchor the good results of change and reform, and to make them robust and sustainable.

6.4 DEVELOPING A WATER LOSS REDUCTION STRATEGY

The water loss reduction strategy will comprise three main themes:-

- Improving data.
- Reducing real losses.
- Reducing apparent losses.

Improving data quality is important as it provides a better understanding of the true level of water losses, but more importantly the different components of NRW and also enables efforts to be targeted in both real and apparent loss components and to geographical areas that provide the greatest benefit. For example, if production or DMA inlet meters are unreliable, they cannot be used as the basis for measuring reductions in leakage that may be used to derive performance incentive payments.

Potential areas of data improvement include the following:

- Recording and reporting, e.g. data, records, assets
- Customer metering, reading and billing systems
- Network monitoring, e.g. pressure zones, district metered areas, system input volume, customer consumption
- Work orders and job recording
- Customer data base

The roadmap to water loss reduction will comprise an action plan aiming to improve data availability and reliability as well as to reduce losses. Potential activities that may be included in an action plan are the following:

- *Billed Authorised Consumption*: Established robust procedures for customer connections both existing and new with properly functioning water meters installed on each customer connection.
- *Unbilled Authorised Consumption*: Water provided for free through registered connections should be minimised as much as possible through discussions with the users, often being municipalities and/or government, on the basis of encouraging water efficiency.
- *Apparent Losses*: Regularly monitor meter readings, test and replace defective meters and hold campaigns to identify and register connections, such as water theft, illegal connections, bypassed meters, etc.
- *Physical Losses*: Establish culture of actively seeking out and rectifying leakage and overflows from storage reservoirs. Apply pressure management to reduce leakage and occurrence of new breaks. Encourage the public to report visible leaks and introduce active leakage control to locate non-reported leaks.

If significant reductions in real (physical) losses are required this will inevitably mean establishing or increasing Active Leakage Control (ALC) effort. Passive leakage control involves repairing leaks that are reported (for example by customers) whilst ALC means proactively looking for and then repairing leaks before they are visible.

The NRW strategy may include improvements to passive leakage control at a utility level, in other words move away from "business as usual". This typically entails establishing a system for reporting leaks (by customers, the public and utility staff), an awareness campaign, and making sure that systems and procedures are in place to effectively repair reported leaks.

The effectiveness of ALC depends on how well leakage detection effort is targeted, the effectiveness of leakage detection and of repairs. It may involve organisational changes for example setting up a dedicated NRW management unit.

A key element of establishing and delivering an effective NRW strategy for the selected areas is the provision of robust and effective means of measurement in different sections of the network such that the Active Leakage Control (ALC) effort can be effectively targeted towards areas of greatest need. The traditional means to achieve this, that has been well established over a period lasting several decades, has been the establishment of a system of District Metered Areas (DMAs), which are discrete areas of the distribution network that are hydraulically isolated from each other and where all inlet and exit points are fitted with a flow meter.

Pipe replacement is another means of reducing losses but needs to be targeted to ensure that schemes are cost-effective. Pressure management, including appropriate rezoning can be an effective means of reducing leakage and bursts.

The key requirement in the short term is to put the necessary structures and systems in place for the water utility to efficiently and effectively manage the network. Experience has also shown that it is not only necessary to structure the technical and financial aspects of NRW but also to assign the responsibility of the execution and a dedicated team has more success than when the responsibility is just added to an existing unit.

Accurate production metering is essential for a reliable water balance. Without a reliable water balance it is not possible to monitor the performance of the operational management in reducing overall Non-Revenue Water. As the tasks of backlog leak detection are implemented it will also be necessary to ensure that there are adequate resources for enhanced leak repair capability – the leak detection activity will locate additional leaks to be repaired.

It is important not only to address real losses but also apparent (commercial) losses focusing on the "Commercial Cycle", which is a chain of 4 sub-processes, namely: meter management, meter reading, customer billing and bill collection and failure in one of these results in poor performance overall. A commercially efficient water utility shall have its commercial activities organized around a seamless commercial cycle that includes the above distinct stages.

- *Meter Management*: includes the meter itself, its accuracy, the activities of meter installation, maintenance, calibration, etc.
  The water utility in order to maintain the highest possible degree of meter accuracy should be in a position to test customer meters both small and large size, determine meter inaccuracies, decide on improvements and maintain relevant records of meters tested, faults and problems encountered using dedicated software and data base. In addition determining suitable specification for all customer meters in order to achieve the maximum measurement accuracy possible under the prevailing water quality conditions as well as recommendations for the proper meter installation should be a top priority.
- *Meter Reading*: entails access to the meter, reliability of the reading, registration practices of the meter reading.
  Meter readings must be checked for errors by the billing department. It is desirable that readings are registered using hand held devices, which minimize transfer errors into the billing system and if possible remote-reading technologies. In addition the GPS location of the meters would

be beneficial to be entered into the GIS for further analysis. It is customary in some cases to use estimated readings, especially for meters in remote areas. This practice is not recommended or at least applied when absolutely necessary and be backed up by real readings in between estimated readings. Reading errors are common and meter readers are not always reliable and measures should be taken to minimize this outcome.

Unauthorized consumption includes illegal connections, meter bypassing, illegal use of fire hydrants, water meter tampering. The main action of the utility towards facing unauthorized consumption is detection, i.e. the ability to become aware of unauthorized consumption in its various forms, as well as enforcement, i.e. the means to halt illegal consumption and invoke appropriate penalties

- *Customer Billing*: comprises data entry of readings, data management (including assumptions, distributing common needs, etc.) and producing and sending the bill tom the customer.

  The customer billing system is an important source of information in water utilities, storing customer account and water meter data as well as customer meter readings from which consumption volumes are calculated. It is important to bear in mind that the value and use of consumption data goes beyond serving as the basis for billings. They are also needed to size service lines, customer meters and water supply infrastructure, to develop hydraulic models and to provide data for non-revenue water management programs. Customer billing systems must therefore be designed to fulfill both billing (financial) and operational (engineering) purposes.
- *Bill Collection*: that is cash collection, possibly the only income many utilities have.

  The collection efficiency should be measured over such period that is representative of the efficiency of collection. In most utilities a certain proportion of the bills are not paid creating the so-called "accounts in arrears". Needless to say that in a situation where consumers receive water intermittently bill collection is a big issue and water utilities are faced with situations of low collection efficiencies. It is to be expected that when people do not receive that required level of service or at least the level of service that they are paying for they will be disgruntled and reluctant to pay.

6.5 WATER SUPPLY MATRIX

High levels of water losses are often accompanied by poor levels of service, particularly an unreliable and intermittent supply to customers. Intermittent water supplies result in a downward spiral of increasing wastage and customer dissatisfaction leading to even higher losses and lower level of service. The goal is to address the intermittent water supply conditions and particularly the water losses, which will help reverse the ever worsening situation and by considering the level of supply in each area of the network, the strategy can be developed to achieve this.

A water supply matrix was developed, Figure 6.6, showing how areas may be categorised into four quadrants, namely:

- High water loss level, intermittent supply
- High water loss level, continuous supply
- Low water loss level, intermittent supply
- Low water loss level, continuous supply

In order to improve operational, commercial and institutional efficiency the water utilities will need to strive towards reducing their water loss level in the first instance coupled with an increase in the hours and days of supply until continuous supply conditions are achieved. A final step in this process once low water loss level with continuous supply is achieved is to reduce and sustain the level of water loss to an economic level.

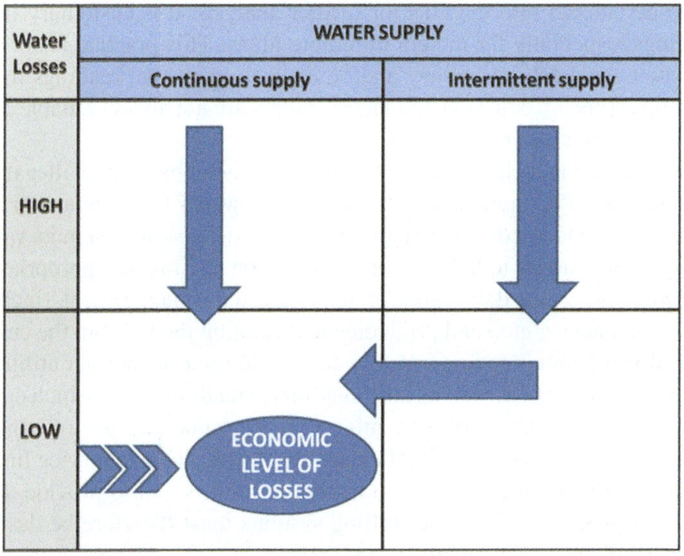

Figure 6.6 Water Supply Matrix.

Priority actions for reducing NRW will be dependent on which quadrant the utility lies. Typical priority actions are summarised as follows:

6.5.1 High water loss level, intermittent supply

The aim in this case is to apply such actions as to move from high to low water losses in order to improve operational performance and management of the network and to build the foundation for moving from Intermittent to continuous supply.

- Assess scope to increase existing supply by repair/optimisation of sources
- Critically evaluate operational conditions and address shortfalls
- Targeted zoning, pressure management and pipe renewals
- Encourage reporting of leaks and leak repairs
- Decide criteria for rehabilitate/replace in asset management
- Pilot Active Leakage control and extend to save water thus moving first area to continuous supply

6.5.2 Low water loss level, intermittent supply

This operational regime is favourable in achieving continuous supply on an area-by-area basis through the implementation of targeted actions as follows:

- Establish continuous supply in a pilot area
- Analyse operational conditions in pilot area
- Determine appropriate action plan based on findings in pilot area
- Gradually up-scale to the whole network (this will inevitably entail actions as Low NRW level, continuous supply)

6.5.3 High water loss level, continuous supply

High levels of water losses under continuous supply conditions are a characteristic of a poorly run system where the cost of producing the water would generally be relatively low. Reduce water losses to a sustainable economic level to ensure that the risk of deteriorating supply is avoided:

- Implement network zoning and pressure management
- Review overall billing policy and commercial database
- Review metering accuracy/policy
- Reduce un-authorized use
- Develop Active Leakage Control
- Encourage reporting of leaks, ensure effective repairs
- Limited pipe replacement based on condition assessment

6.5.4 Low water loss level, continuous supply

Water utilities that have achieved a low water loss level with continuous supply are well on their way to reach a sustainable level of water losses at an economic level. Maintain at an economic level and ensure that the supply/demand balance is maintained and improved efficiency is achieved:

- Reduce customer meter and accounting errors
- Improve active leakage control
- Improve repairs (materials, workmanship, response time)
- Maximise pressure management
- Develop demand management programmes
- Review tariffs if needed

6.6 PREPARING AN "IWS TO 24 × 7" TRANSITIONING PLAN

6.6.1 Introduction

The challenge of managing a water supply system that is intermittent and with high levels of NRW poses water utility managers with a conundrum regarding how to address these. Although there is a generic approach for quantification and management of NRW based on IWA best practices, their application to intermittent water supply systems poses great challenges to practitioners.

Before the first edition the IWA manual on Performance Indicators (Water Impact Guidebook) was published, there was no international attempt to standardize the water balance and water loss performance indicators. The IWA water balance and water loss PIs have meanwhile become international standard and are promoted by many regional and national professional associations around the world (including AWWA).

It is well known that expressing water losses (or NRW) in percentage of system input is misleading in the best case and does not work at all in an IWS situation (No wonder that % water loss can be low if a utility has only a few hours water supply per day).

Water loss performance indicators, for example physical losses in litres/connection/day, always need to be adjusted to continuous supply (the acronym used is "w.s.p." – when the "system is pressurized").

For example: When in a system with 10,000 service connections and IWS of 4h/day physical losses are 3,000 m^3/d the correct performance indicator would be:

- 3,000 m^3/d/10,000 connections = 0.3 m^3/conn./d (300 l/conn./d)
- 300 l/conn./d 4h × 24h = 1,800 l/conn./d (w.s.p.)

Only with this indicator (and the average operating pressure) the level of water loss can be understood and the transformation from IWS to 24 × 7 planned.

In summary, the IWA water balance methodology and the IWA water loss PIs can also be used in IWS systems – IF the supply time is properly taken into account.

Once the water loss situation is properly understood, forecasts can be made how much water will be required to supply the network in its present condition on a 24 × 7 basis and how much will be needed after network rehabilitation.

Transitioning from IWS to 24 × 7 will be different depending on the type of IWS:

- If the system was designed for IWS (like most in South Asia) one needs to start with pressurizing the system 24 × 7 on a zone by zone or DMA by DMA basis starting from the zone or DMA closer to the water source.
- In systems where IWS was not planned but became a reality in fringe areas of the system, water loss reduction (again, zone by zone) must be started in the part of the network with best supply and highest water losses (normally the areas which still receive 24 × 7 supply) and the water saved can then be pushed to the poorly supplied areas. It is often difficult to argue why interventions are needed in the well supplied area while people have the feeling that works should be focused on areas with IWS.

The World Bank is advocating that transitioning from IWS to 24 × 7 supply should be done in a cost effective manner combining improved operations with targeted capital works. This approach has to be based on sound data, good technical modeling, the introduction of good management practices, particularly related to finding and fixing leaks and the assessment of different options to find the most cost effective solution. Therefore the approach for a successful and sustainable transition from IWS to 24 × 7 supply should be two-fold:

(1) To move from intermittent supplies to 24 × 7 in a cost effective manner – by improved operations and maintenance, by restructuring the network to improve hydraulic performance, by replacing part of the distribution system as needed, and by introducing Active Leakage Control to reduce losses in a cost effective manner.
(2) To create a well-functioning utility that has the capacity to continue to operate the system at the end of the PPP contract – by using the PPP to build that utility through introduction of modern management systems, procedures and equipment.

In the next sections it is attempted to draft the outline of a future standard approach for transitioning from IWS to 24 × 7.

6.6.1.1 Step 1: IWS problem analysis

The first step is to analyze why there is IWS and to what extent.

Questions to be asked are:

- **Was the system designed for IWS?** This means that it was always planned to bring water only once or twice per day to the consumers? In these cases you normally have oversized pumping stations and trunk mains since the same volume of water needs to be supplied during a short period and therefore the flow rates during this period are high. In such systems most, except the urban poor, customers have ground and roof tanks. As soon as the supply is on all ground tanks get filled and therefore pressure in the pipes is very low.

- **Has IWS been introduced recently in system that was previously supplied on a continuous basis?** This would typical for an unexpected drought period where the utility's management is afraid that they may run out of water and hope that with IWS water demand can be substantially reduced. In this case most customers won't have ground tanks and pressure will be normal during supply hours. But pressure peaks when opening and closing to schedule supply to different areas (or the entire system) will cause extreme pressure fluctuations and transients, which will have a massive impact on the burst frequency.
- **Has the system been always supplied 24 × 7 but nowadays some areas only receive intermittent supply?** This means is it the intention to supply the entire system 24 × 7, the pumps are continuously operated but some areas, which maybe in the past did have continuous supply, receive water on an intermittent basis? Such systems, especially close to the concerned areas, are characterised by extremely low pressures during peak demand hours and high night-time pressures. Usually only during these night hours water will reach the IWS areas.

Transitioning to 24 × 7 will need different strategies for the different situations. The proposed methodology, (Step 2–5) is suitable for the first and second scenario outlined above. In the third case a hydraulic analysis of the distribution network is required, hydraulic bottlenecks have to be detected and areas with highest levels of physical losses (normally areas that are still supplied 24 × 7) have to be identified. In these areas physical loss reduction has to start and water saved can then be supplied to other areas which at present experience IWS.

6.6.1.2 Step 2: Initial water audit

Ideally a full water audit would be carried out resulting in an IWA Water Balance with 95% confidence limits.

But if time, funding, or expertise for a full water audit is not available, at least an initial audit, or "Rapid NRW Assessment" is required. It is a simple, fast and usually inexpensive methodology to get a first understanding of the NRW situation in a water utility. It is a precondition for understanding the magnitude of the IWS problems and the possibilities of transitioning to 24 × 7.

(1) Collect all necessary information required to determine the following:
  (a) Daily volume of NRW
  (b) Average supply time (in case of intermittent supply)
  (c) Average pressure
  (d) Number of service connections.
(2) Determine the annual average daily volume of NRW. Review available records and identify possible irregularities and determine the most likely average daily volume. Review billed consumption data, both billed metered and billed unmetered consumption and determine the annual daily average volume of billed consumption.
(3) Calculate the average daily supply time. Identify areas with different supply times, determine the approximate number of service connections of each area and calculate a weighted average supply time using the number of connections as weighting factor.
(4) The average pressure in a certain point is the 24 hour average. Pressures will be recorded with electronic pressure loggers for at least a 24h period at each point. If no pressure loggers are available, pressures during low-pressure periods (normally during peak demand time) and during high pressure periods (usually night time pressures) shall be measured with pressure gauges. Identify areas with different pressure characteristics, measure pressures and calculate average pressure for each area, determine the approximate number of service connections of each area and

calculate a weighted average pressure using the number of connections as weighting factor. Please note that only pressure during supply hours must be used in the calculation.

(5) Determine the number of service connections (SC), which is in many cases not equal to the number of customers. Only in cases where each customer (account) is supplied by an individual service connection the number of connections equals the number of customers. If, for example, every single condominium in a building is considered one customer (=one account) then the number of service connections will be significantly less than the number of customers (one condominium building may consist of, for example, 100 customers but is supplied by only one service connection.

This seems all straight forward. But what to do if there if the volume of NRW is not known?

(1) **System Input Volume**: If the system input volume is not metered, short time measurements with portable flow meters are required. It is important to get at least crude data on the average daily system input volume.
(2) **Consumption**: In some water utilities, especially in South Asia, consumption is not metered and consumers are either not billed at all, or billed on estimates or the value of their property. It is of extreme importance to get an understanding of the actual water consumption. There is no other way than installing water meters, ideally high accuracy meters with data loggers, at about 100 truly randomly selected domestic customers. Customers must be metered for at least one week. Then either use the average consumption (per customer/day and capita/day). The consumption of non-residential customers (commercial, institutional, industrial) must be assessed separately. Once this is done, you can calculate the total water consumption (whether billed or unbilled) of the area.

6.6.1.3 Step 3: Water loss performance indicators

After completion of the water audit or the Rapid NRW Assessment the physical loss performance indicators need to be calculated.

Physical losses will be expressed in liters per service connection per day per meter (head) average pressure and adjusted for the average daily supply time:

- l/conn./d/m (w.s.p.)[2]

Below the step by step instruction:

(1) Determine daily volume of Physical losses. These are the different scenarios:
    - Water audit available – simply take average daily volume of physical losses from the Water Balance
    - Metered consumption data available but no water balance:
        - assume volume of physical losses = 90% of the volume of NRW
    - Consumption data not available and consumption measurements have been carried out:
        - [Physical losses (PL)] = [system input volume] – [calculated volume of water consumed]
(2) Divide average daily physical losses (PL) by the number of service connections and express result in liters/connection/day (PL in l/conn./d)
(3) Adjust to supply time:
    - [PL l/conn./d]/[average supply time (h/d)] × [24h] = [PL l/conn./d (w.s.p.)]
(4) Divide by average pressure:
    - [PL l/conn./d (w.s.p.)]/[average pressure (meter head)] = [PL l/conn./d/m (w.s.p.)]

[2]w.s.p. stands for "When the System is Pressurized" and means that the value of a performance indicator is adjusted (extrapolated) to the level it would have under continuous (24 × 7) supply conditions.

6.6.1.4 Step 4: Volumetric assessment

This is one of the most important, and often overlooked, step in planning the transitioning from IWS to 24 × 7.

First it needs to be calculated how much additional water (system input) would be required to supply the area under consideration 24 × 7 with the network still in its present, leaky, condition. In a second step a target leakage level for the rehabilitated network has to be determined and as the third step the volume of physical losses and then the total water demand for 24 × 7 and a rehabilitated network can be calculated. In detail:

(1) Additional volume of water required:
  (a) Volume of physical losses for present network condition and 24 × 7 supply:
    [Number of SC] × [PL l/conn./d/m (w.s.p.)] × [expected average pressure (meter head)]
    Note: it needs to be expected that pressure will increase when supplied 24 × 7; make a reasonable assumption based on the local situation
  (b) [Volume of PL for 24 × 7] – [Volume of PL for IWS] = [Volume of additional water required]
(2) The target leakage level will depend on the present leakage level. As a first suggestion, the following formula might be used to get an idea about a realistically achievable physical loss target level:
    [target PL l/conn./d/m (w.s.p.)] = [present PL l/conn./d/m (w.s.p.)] × 0.05 + 8
    Once this (still high) level of physical losses has been achieved further physical loss reduction programs can be designed and implemented to reduce physical losses further until the most economic level is reached.
3. Calculation of the future system input volume at 24 × 7 supply, target level of physical losses and target pressure:
  (a) Volume of physical losses for future network condition and 24 × 7 supply:
    [Number of SC] × [target PL l/conn./d/m (w.s.p.)] × [target average pressure (meter head)]
    Note: the target pressure will depend on the hydraulic situation and the level of service policy of the water utility
  (b) [Future Volume of PL] + [Consumption] = [Future System Input Volume]

6.6.1.5 Step 5: Planning the transitioning from IWS to 24 × 7

The last step is to decide whether there is enough water available to supply. Based on the volumetric assessment carried out as outlined in Chapter 6.6.1.4 (Step 4: Volumetric assessment), it can now be determined whether there is enough water available to immediately supply the entire system or the target zone 24 × 7. In general this will not be the case else there would be no reason for IWS. Therefore small zones or even DMAs have to be established, close to a point where 24 × 7 supply is possible while the rest of the distribution network can still be operated intermittently. This will normally be close to a reservoir, treatment plant or along a transmission main, which is always pressurized.

Once physical losses have been reduced the water saved can be used to supply the next isolated zone and so on.

Transitioning from IWS to 24 × 7 supply is not an easy task. It requires commitment and dedication from all concerned, governments, water service providers and consumers. The following areas need to be addressed simultaneously and in parallel in order to have a successful and sustainable impact:

- Technical – gradual increase in the hours of supply aiming for continuous service, introduction of customer metering policies, improved network operation using DMA/sectorisation practices and targeted rehabilitation/replacement of mains.

- Financial – implementation of tariff structures linked to performance incentives for saving water, cost recovery, adoption of commercial thinking and reform of water service providers to make them accountable.
- Institutional – water service providers that have fallen into Intermittent Water Supply have major governance and incentive flaws and need in-depth reform; moving to continuous supply requires often very difficult political and institutional choices that many governments prove reluctant to make. A paradigm shift is imperative!
- Social – the water service providers need to gain the trust of the consumers, have the willingness to change, involve the public in this effort in order to provide the required level of service at all times in a reliable and sustainable manner.
- Communication with the customers – It is of the utmost importance to communicate all the above in an effective and convincing manner to all involved in order to have the maximum possible impact.

Transitioning from IWS to 24 × 7 supply is possible, applying the appropriate approach, techniques, methodologies and practices. However, it requires commitment and dedication from all concerned, governments, regulators, water service providers and consumers.

A number of pilot projects have been executed over the past few years in India aiming at increasing the hours of supply to the consumers and ultimately supplying 24 × 7 service. Karnataka Urban Water Sector Improvement Project (KUWASIP) implemented a 24 × 7 demonstration project via Public Private Partnership (PPP) in three cities, Hubli–Dharwad, Belgaum and Gulbarga, with the aim to achieve 24 × 7 water supply. All three cities have very similar characteristics in terms of water supply infrastructure and assets with water supply and distribution networks mostly constructed of PVC pipes. The supply conditions vary from about 3 hours every 2 days to 3 hours every three days.

The project was funded by the World Bank and a total of 5 Demo Zones were created in the three cities and commissioned in 2006. This work involved replacing almost the entire existing network in the Demo Zones, including customer service pipes, with brand new polyethylene pipes. All customers were given a metered supply, which improved enormously the level of service. This project was able to transition to 24 × 7 with a "big bang" approach – based on replacement of the complete distribution system under the Project, which of course delivered 24 × 7 supply.

6.7 REFERENCE

Charalambous B. (2005). Experiences in DMA Design at the Water Board of Lemesos, Cyprus. In: Proceedings of IWA Special Conferences 'Leakage 2005', Halifax, Nova Scotia, Canada, 12–14 September 2005.

Chapter 7

Leakage detection under IWS

Stuart Hamilton

There are many difficulties when conducting Active Leakage Control in areas of Intermittent Supply and many parameters that ought to be considered prior to starting any leakage activities. To illustrate this set out below are some simple case studies highlighting issues that could be experienced and should be considered when dealing with leakage under IWS.

7.1 CASE STUDIES

7.1.1 Middle East 1

An area in this Middle Eastern city was on intermittent supply and there was a plan to move the area to 24 × 7 supply and to conduct leakage detection. The area was supplied with water for 4 hours per day every 3 days. The consumers used roof tanks to store water, as is the case in most intermittent supply situations. The supply was erratic and the consumers had no ball valves in their roof tanks in order to ensure that they had their water tanks full when the supply was turned on and to have a visual indication when the tanks were full by observing the overflow from the roof tank.

In such situations, where consumer roof tanks have no ball valves any attempt in performing leakage detection would be fruitless as there would be continuous flow through the customer service connections acting as "potential" leakage points. In addition it would be difficult to build pressure since the overflows from the consumer tanks would act as open ends in the network. It was therefore decided as a first important step to install ball valves in all customer roof tanks in order to control overflows and wastage.

The installation of ball valves to the consumers' tanks was initiated by the water utility without prior consultation with the consumers to inform them of the purpose of such action and of the benefits both for them and the water utility. As a result most of the consumers were upset and reacted angrily and removed taps and ball valves installed by the utility and disposed these in the streets.

Bambos Charalambous and Chrysi Laspidou
doi: 10.2166/9781780407074_057

The customers protested and argued with the utility that for many years they were accustomed to their tanks overflowing (Figure 7.1) which for them was a real indication that their tanks were full. To install ball valves in their tanks the customers felt that they wouldn't have the usual visual indication that their tanks were full and this have made them nervous, disgruntled and wanted to go back to the old ways. Following this turn of events it was decided to postpone the leakage detection activities until after a consumer awareness and educational campaign program was introduced showing the benefits of controlling water losses and the type of control devices used to do this as well as the benefits of transitioning from IWS to 24 × 7 supply. The project continued some months later allowing leakage detection activities to take place. It is therefore strongly recommended that communication with consumers is of the utmost importance and must take place early on in the project informing the consumers of the changes that will take place in the water supply regime and the benefits that these changes will bring specially to them and to the utility in general.

Figure 7.1 Customer storage tank without a control ball valve overflowing.

7.1.2 Middle East 2

This case study had a very similar water supply situation as in Case Study 1 but in this instance a different approach was adopted for moving forward with an Active Leakage Control (ALC) program. Instead of installing ball valves to customer roof tanks it was decided to manage the situation in a much different way and to temporarily isolate the supply to customers in order to deal with leakage in the mains and service connections. To do this all customer service connections were located and turned off on a day that the consumers were not to be supplied with water. The mains distribution system was pressurized allowing mains side leakage activities to take place and to locate leaks adopting a traditional approach for repair by the maintenance crews. The steps followed are set out below and should be adopted when working in a system that was on IWS and moved to 24 × 7:

- Conduct walk route of mains to look for any obvious leaks from fittings, connections or leaks that are now showing on the surface (Figure 7.2),
- Conduct an acoustic survey preferably with an electronic listening device but can be completed with a traditional listening stick and record areas of interest for follow up.

- Conduct a correlation survey to check areas of interest and to narrow down to smaller areas to use the listening device to pinpoint the position of leaks. If a correlator is not available then the surface above the pipe should be listened on every 1 metre listening for sound that would be transferred to the surface from the leak.

Figure 7.2 Leak on a service connection identified through a visual inspection of the system.

7.1.3 Middle East 3

In this case study the city was located in an extremely hot geographical area and water was scarce with intermittent regime being the normal supply situation. The mains and service connections were in relatively good condition and customer service connection to each property was fitted with a working isolating valve. Despite the network being in a good condition conducting an acoustic leakage survey was deemed to be impossible under the intermittent supply regime so the alternative of gas injection was used into the dry (empty) water mains. The process was such that all property connections were turned off and areas of the system under test were enclosed to control the area of the network worked in. Gas with a very low level of Hydrogen (welding gas could be used) was injected along with air from a compressor and the process to find leaks using this method was conducted. The methodology used as follows:

- The area to check for leaks was isolated to a length of pipe that could be managed
- The service connections were isolated
- The gas was injected into the main via an existing connection such as a fire hydrant or through a service connection
- A fitting at the end of the pipe was opened and a 'gas sniffer' device was held over the opened fitting to register gas presence
- The fitting was closed when gas presence was registered by the "gas sniffer" device indicating that the main was full of gas
- The 'sniffer' was placed on the ground every 1 metre maximum apart and wait for 10 seconds

- If gas was present an alarm was activated and the only way it made its way to the surface was via a hole or crack of some sort in the water main
- Gas will permeate through all materials including concrete
- Gas is expensive so it needs to be controlled entering the pipe and only use the smallest amount necessary
- Gas can be injected into a wet main and use the water as the carrier of the gas to the leak or if in a dry main then a compressor can be used to mix the gas with air – using gas alone would take large quantities
- There is no restriction to size of pipe or material when using gas

The process was successful in finding several leaks however it was not suggested that this would find every leak. This is a successful method of finding leaks in a dry system.

7.1.4 Asia

The city was supplied by a very low-pressure system, less than 5 m pressure, and it was practically impossible to locate leaks with acoustic methods. What was tried with success in these areas was to locate areas of leakage using a step test locating the street that had the problem and then isolate the street from the system. With the use of a tanker lorry water was pumped into the main via a fire hydrant in order to pressurise the main in the particular street and to enable the leak(s) to be located via acoustic methods. This was an extreme and difficult exercise but did allow the leak to be located with success. In areas of a network that has not been moved to 24×7 this approach could be used if water for ALC is not sufficient to do the whole area. The main priority is to control the area, bring it down to a street level if possible, and not to try to use this approach in an open system.

It should be noted that such an approach is not practical on a larger scale, greater than a street level, as the volumes of water required would be prohibitive and the procedure complex and cumbersome. In addition turning off service connections may be problematic as ball valves (stopcocks) may only be installed on some of the connections or perhaps not installed at all in which case there is an added complication to the whole process. Furthermore the temporary increase in pressure in order to find the potential leaks may in fact cause additional leaks thus worsening the situation. Therefore care must be taken in pressurising the system maintaining at all times a close monitoring of water input into the main and controlling of the pressure.

7.1.5 Africa

Pressure in this city was low and the system was relatively old with very few house connections visible. It was decided that a visual leakage survey should be conducted in the first instance. The visual survey was conducted during the time that the network was supplied with water and relatively pressurised. This simple method identified numerous leaks from mains and service connections as well as leakage from open pipe ends. On average it was calculated that a leak was located at every 100 m of main pipes. It is highly recommended that this procedure, visual survey of the network, is adopted in most cases as a first step before climbing the ladder of leakage identification and location prior to the introduction of any technology.

This procedure is simple, inexpensive and can be conducted with limited training and will identify many leaks.

Process of work is as below:

- Identify area that has water and is pressurized
- Identify a map and plan a walk route
- Walk the main checking for any visual leaks

- Check all house hold connections checking for leaks up to the meter of boundary
- Check the household for any internal leaks present
- Record all findings on a sheet and present for repair or to be dealt with using the company procedure (Figure 7.3)

Figure 7.3 Walking the planned walk route.

7.1.6 Conclusion

It should be understood that when conducting leakage activities in intermittent supplies using acoustic methods it is imperative that the system must have some pressure and the higher the better. For mains that are shallow then a low pressure of 5 metres will be sufficient to show water on the surface in many instances but on mains that are deep and are buried at least 1 meter deep then a pressure of at leaks 10 metres is desirable. This pressure will let the energy from the leak point along the pipeline transmit to the receiving acoustic equipment. All property service connections must be turned off or stop drawing water into the property otherwise it would interfere with the acoustic process indication showing that each property has a potential leak on it and would have to be investigated hence causing Active Leakage Control to be long and non-productive process. The water supply has to be in the system long enough to fill all customers tanks and each tank has to be controlled via a ball valve or other mechanism so that when full there is no flow through the customer service connection.

The method of gas injection is an alternative to the acoustic methods but it is more costly, requires a significant amount of gas to cover a system, is not readily available and requires experienced operators. Usually this service is provided by specialised contractors, who have the appropriate equipment, technology and knowhow.

The alternative of visual leakage activity in reality requires no experience and no technology and is probably always the 1st approach to be adopted and should only be carried out when the water is turned on as described in Case Study 5 above.

7.2 METHODOLOGY FOR LEAKAGE DETECTION ACTIVITIES

This approach is applicable for all types of activities regardless of the supply mode, IWS or 24 × 7, and should always be considered when planning Active Leakage Control activities. It is therefore absolutely essential that before heading out into the field, all available information for each network area should be collected and analysed on which the choice of appropriate leak detection methodology will be based, taking into account certain site characteristics such as:

- Location of the area (urban/rural)
- Length of mains
- Customer type
- Number of service connections
- Pipe and service connection material
- Average pressure
- Asset condition
- Hours of supply

7.2.1 Important issues

Characteristics: *The area.* Understand in advance the geographic area where the crew will be working and in what surrounding; what the time of day is that the work will be completed; and what is required to complete the work safety-wise. This will affect what leak detection approach should be adopted and what technology should be used. Understanding this at the outset will avoid going blind onto a site, and thus the crew will be prepared for the local situation that they will encounter.

Characteristics: *Mains length.* Determine in advance the length of mains in the area, as this will determine how long it will take to complete the leak detection survey, and thus the time scale it will take to complete the area as a whole.

Characteristics: *Customer type.* Gather information in advance on the characteristics of the customers where the leak detection survey will be conducted.

- Is the customer type commercial or domestic?
- Is the housing large or small, apartments or houses?
- Is the area wealthy or poor?

This information helps to properly plan the leak detection approach needed prior to going onsite. If the area is commercial, conduct meter readings on large customers prior to completing the leakage surveys. Note any large customers in the middle of a domestic region (such as hospitals or hotels) that may consume large quantities of water 24 hours a day. Noting in advance these large customers as causes of high usage will prevent them from being mistakenly considered leakage.

Characteristics: *Number of service connections.* Determine the number of customers and service connections in the area, if possible using GIS data. If not included in the GIS yet, use meter reading books. This is important in order to determine the number of fittings available to listen on.

Characteristics: *Pipe material.* Knowing in advance the pipe material content along with the pressure is critical, as there is a different approach to leakage in metallic and non-metallic regions.

Characteristics: *Average pressure.* The choice of leak detection technique and the expected difficulties largely depend on the operating pressure. In general, the lower the pressure the more difficult it is to find leaks.

Characteristics: *Asset Condition.* Obtain information on the condition of the network asset in the area to be surveyed, such as age, number of prior repairs, and any other information that might be available.

Characteristics: ***Hours of supply.*** Obtain information on the operational time that the water is into the system bearing in mind that only during this period can leakage activities take place. For acoustic leakage activities the system has to be controlled to obtain highest pressures while for visual activities this can only be carried out during the hours of supply.

A successful leak detection plan requires knowledge of the above characteristics of an area to enable proper planning in the office prior to going onsite. This information will determine the leakage location activity, and the length of time to conduct the survey based on the number of connections per day and or length of mains is in the area. Knowing the distance between fittings and the pipe material, along with pressure, will also help to determine what approach to undertake and what technology will work best. Knowing the housing/dwelling type will allow you to consider consumption per property, which will assist in working out whether the problem is high usage or leakage. Also, knowing the extent of commercial activity in the area will dictate as to when to conduct the leak surveys.

In case water availability into an area under intermittent supply conditions is during periods of high noise or traffic it should be examined to temporarily change the supply regime to a quiet period which normally is during the night time in order to be able to conduct leakage detection activities under better conditions.

7.2.2 Further considerations and recommendations

Urban or Rural? If the area is rural with large length of mains, but with few properties, conduct a Step Test rather than walking the mains. Likewise, if the area is urban with heavily populated properties and significant traffic noise, then nighttime leakage detection may be the better solution. However, before deciding to work at night, conduct a daytime visual inspection to understand where the fittings are and where the water mains run.

Pipe material: ***Mainly metallic.*** If the area is mainly metallic, consider the distance between fittings. Acoustic noise will travel further in a metallic system, allowing the noise to be heard hundreds of meters away. Acoustic technology such as correlators will work easier in these circumstances.

Pipe material: ***Mainly non-metallic.*** In non-metallic areas, listening on fittings will not work. The noise of any leak will not be easily heard since the energy created by the leak is absorbed by the non-metallic material. In these circumstances, acoustic devices such as correlators are of very limited use, and other approaches need to be considered. Top sounding for leaks is one of the suitable approaches.

Confirm security of existing boundary. In the case of DMAs, ensure that all boundary valves are in a closed position and that no water is leaving the area to be surveyed. This is important, as the indication of high leakage may be from a valve being open and not from an actual leak.

Locate & sound all mains fittings. Locating all fittings during daylight hours is imperative, as many hours can be wasted at night looking for a valve to conduct a listening survey or step test. Here is where advance preparations are particularly critical.

Deploy ALC plan for urban region. Understanding the type of region will determine which methodology will provide the best chance of quickly locating the problem. The sooner a problem is located, the sooner it can be rectified. Urban region characteristics such as heavy traffic, high population, and proximity to schools and commercial businesses will mean that the time of day to conduct leak surveys must be planned in advance.

Deploy ALC plan for rural region. In rural regions, fittings are few and the distance to survey is high. So, either a drive through the area as a first instance or conducting a step test to narrow down the problem should be considered.

Verify areas of interest. Not all findings will be leaks, so initially these should be classified as "areas of interest" (AOI) where the technician must verify by all possible means that the AOI is actually a leak.

Because of the high cost and time frame to excavate, "dry holes" or aborted excavations must be avoided and costs incurred when the problem is not located.

Target. Determine the volume of leakage to locate, and estimate the volume of the individual leaks that have already been located. This is necessary to determine if the target of the leakage volume has been reached. Consider withdrawing from the region until the leaks have been repaired and a new measurement taken. If the exit target has been reached, then move on to another area.

Develop and implement quiet hours "step testing" plan. Produce a step test plan to carry out the valve closing exercise at a time when most convenient, when consumption is at the lowest in the region. Not all step tests have to be completed at night. Many can be completed during the day such as in rural areas.

7.2.3 Trunk main leak detection

Leak detection on large diameter pipes is difficult and the following activities should be considered when doing leak detection on trunk mains:

- Inspect all valve chambers (including air valves, washouts, control valves, hydrants, etc.) for signs of leakage from valve glands, pipe joints etc.
- If leakage is suspected, sound the pipeline as a whole, using a ground microphone.
- Examine discharge pipe work from washouts along each pipeline.
- Inspect adjacent ditches, dikes and field drain discharges, for signs of leakage from the pipeline.
- Inspect all meter installations at the start, along and at the end of each pipeline.
- Take flow readings for use in carrying out a Water Balance for each transmission main system.
- Identify and confirm all branches and connections (including service connections) from the pipeline, and assess the magnitude of the flows through these connections.
- Ensure that minimum damage is caused to any crops, hedges, fences etc. when crossing private lands, and that all gates are left in the same position as on entry.
- Immediately report any matters of consequence arising from entry onto private land to the supervisor.

7.3 SCENARIOS FOR LEAKAGE DETECTION

The following typical scenario examples are provided to demonstrate the choice of appropriate leak detection methods and the steps to be taken in each method. The methods presented are derived from previous experience working in similar utility networks. The methods assume both day and quiet hours' operation.

7.3.1 Scenario A: IWS regime of at least 4 hours and higher

Urban Area of 1,500 service connections; 5 kilometres of mains length consisting of mixed material; a majority of domestic properties and a small number of commercial properties; having residential streets and main roads; and operating pressure of below 10 m.

Method Steps:

- Determine the volume of leakage in the area based on the number of hours of supply
- Obtain a GIS plot of the area, incorporating the boundary and associated normally-closed valves if applicable
- Check records from recent leakage management activity from past surveys.

ALC step 1

- Complete a sweep of the area by visual leakage only to find the simple and easy leaks where no experience or technology is required.
- Investigate visual survey for water running into drains or sewer channels

ALC step 2

- Carry out initial leak noise location through sounding all services using the electronic listening device when the system is pressurized, to locate the "area of interest."
- Listen at every metre interval directly above the mains pipework during lowest noise and highest pressure periods
- Pinpoint the leak location and prepare a leak location report.
- Read meters of commercial properties to ensure correct allowances are made in the water balance calculations
- Re-evaluate the level of leakage, taking into consideration the estimated losses from the leaks located. If results are presumed to be insufficient, continue surveying the region until the estimated number of leaks has been found and the target leakage level is likely to having been achieved.

ALC step 3

If no more leaks are located using traditional acoustic methods, then the following procedures should be used:

- Locate and sound all fittings within the zone including every service connection and record the noise level and plot onto a map to show any potential 'noise clustering'
- Investigate reducing the area of the zone
- Consider installing valves or temporary disconnect mains and measure flows in the smaller zonal areas
- Locate areas of high losses, and concentrate leakage activity into these areas
- Investigate illegal connections in region as cause of high level of losses
- Investigate increasing pressure and conduct survey with electronic listening device during this period
- Continue to survey the area during quiet periods along the main
- Investigate visual survey for water running into drains or sewer channels
- Investigate possible unknown commercial properties or domestic properties carrying out commercial activities
- Continue with the above strategy until targets have been achieved.

ALC step 3 (possible alternatives)

- Consider Gas injection
- Consider isolation of system and pressurize small areas with tanker and pump
- Consider internal inspection technology

Ensure that the ALC work is carried out when it is deemed the supply has filled the consumer tanks so they will not be misinterpreted as leakage.

7.3.2 Scenario B: IWS regime to temporarily 24 hours in an urban environment

In this scenario in order to allow ALC activities to take place in an urban area under IWS the area will temporarily be moved to 24 hour supply to provide the necessary conditions for leakage detection.

Urban area of 1,500 service connections; 10 kilometres of mains length consisting mainly of metallic pipes; a majority of domestic properties and a small number of commercial properties; residential streets and main roads with a 24 hour water supply regime; and operating pressure of at least 10 m.

Method Steps:

- Determine the volume of leakage in the area.
- Obtain a GIS plot of the area, incorporating the boundary and associated normally closed valves.
- Check records from recent leakage management activity from past surveys if any.
- Carry out initial leak noise location through sounding & correlation techniques, to locate "areas of interest."
- Pinpoint the leak location and prepare a leak location report.
- Re-evaluate the level of leakage, taking into consideration the estimated losses from the leaks located. If results are presumed to be insufficient, continue surveying the region until the estimated number of leaks has been found and the target leakage level is likely to having been achieved.

If no more leaks can be found using acoustic methods, then the following procedures should be used:

- Locate and sound all fittings within the zone.
- Develop a "step test" strategy
- Carry out the "step test" strategy
- Locate areas of high losses, and concentrate leakage activity into these areas.
- Continue with the above strategy until targets have been achieved.

It is extremely important to ensure that the ALC work is carried out several hours after the water supply into the area has been resumed so that sufficient time is allowed for consumer tanks to fill-up and the possibility of filling of tanks filling during the ALC activities is minimal thus avoiding misinterpretation of water taking by consumers a as leakage.will not be misinterpreted as leakage.

7.3.3 Scenario C: IWS regime to temporarily 24 hours in a rural environment

In this scenario in order to allow ALC activities to take place in a rural area under IWS the area will temporarily be moved to 24 hour supply to provide the necessary conditions for leakage detection.

Rural Area of 200 service connections; 30 kilometres of mains length consisting of mixed mains material; and operating pressures not less than 10 m.

Method Steps:

- Ascertain the current level of leakage from existing area metering, and provide data for entry and exit.
- Obtain a GIS plot of the area incorporating the boundary and associated normally-closed valves.
- Check records from recent leakage management activity from past surveys if any.
- Develop a "step test" strategy.
- Carry out the "step test" strategy.
- Locate areas of high losses, and concentrate leakage activity into these areas.
- Carry out leak noise location through sounding and correlation techniques in "areas of interest."
- Pinpoint leak location and prepare leak repair report.

- Re-evaluate the level of leakage, taking into consideration the estimated losses from the leaks located. If results are presumed to be insufficient, continue surveying the region until the estimated number of leaks has been found and the target leakage level is likely to having been achieved.
- If no more leaks can be found using acoustic means, revert back to another step test.
- Continue with the above strategy until exit levels have been achieved.

It is extremely important to ensure that the ALC work is carried out several hours after the water supply into the area has been resumed so that sufficient time is allowed for consumer tanks to fill-up and the possibility of tanks filling during the ALC activities is minimal thus avoiding misinterpretation of water taking by consumers a as leakage.

Chapter 8

Metering under IWS

Dimitris Foufeas and Nicholas Petroulias

Intermittent water flow in supply and distribution networks may result in the deterioration of water meter accuracy ultimately leading to malfunction due to high water velocities developed during the filling of the pipelines, the transport of suspended solids in the system and expulsion of air through the meters thus seriously damaging the registration mechanism. This situation affects most of the water meters installed in supply networks whether they are used to measure bulk water flows or customer consumption.

8.1 DOMESTIC WATER METERS

The operating principle of domestic water meters installed in the vast majority of water supply distribution networks is based on the mechanical transmission through a moving mechanism located in the water stream to a volume totalizer (Figure 8.1). These types of meters, referred to as *Mechanical Meters*, measure the velocity of flow through a meter of a known internal volume. The speed of the flow can then be converted into volume of flow to determine the usage. There are several types of meters that measure water flow velocity, including jet meters (single-jet and multi-jet), turbine meters, propeller meters and mag meters.

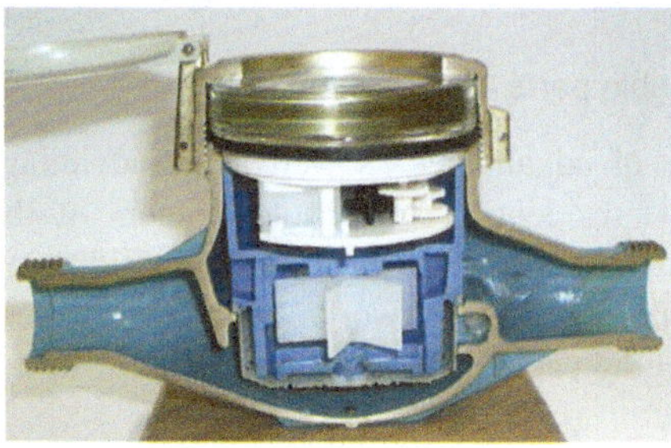

Figure 8.1 Cross section of a velocity type meter.

Bambos Charalambous and Chrysi Laspidou
doi: 10.2166/9781780407074_069

The volumetric type meters (Figure 8.2), often referred to as *Positive Displacement Meters*, rely on the water to physically displace the moving measuring element in the water meter in direct proportion to the amount of water that passes through the meter. There are two typical types of Positive Displacement meters, oscillating piston meters and nutating disk. The piston or disk moves a magnet that drives the register to record usage. Either method relies on the water to physically displace the moving measuring element in direct proportion to the amount of water that passes through the meter.

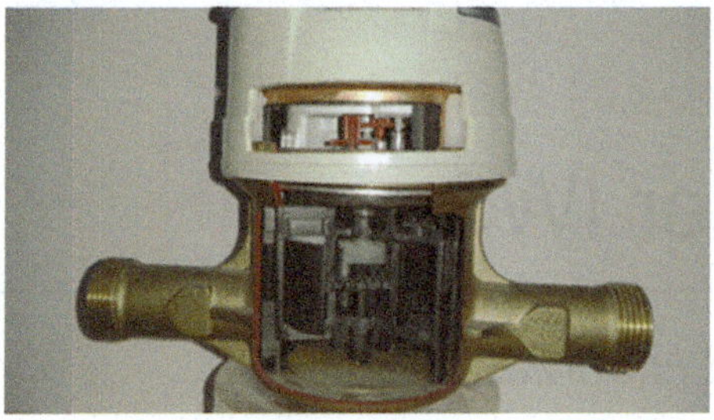

Figure 8.2 Cross section of a volumetric type meter.

In the last few years a new technology in manufacturing of domestic meters has emerged which has no moving internal mechanism for flow measurement exhibiting excellent long-term stability and reliability as well as being extremely sensitive and accurate in low flows.

The principle of operation of this new generation of water meters is relying mostly on the ultrasonic and/or the electromagnetic technology and more recently using residual magnetic field technology. Because of the use of advanced electronics in manufacturing these meters, they are equipped with multiple advanced features and operations, and are therefore often referred to as *Smart Meters* (Figure 8.3).

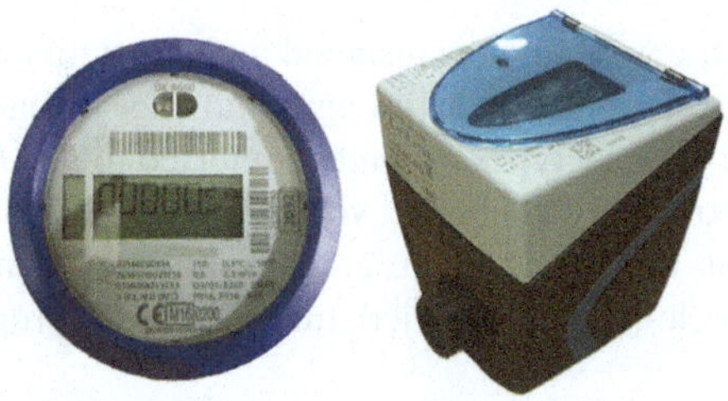

Figure 8.3 Meters with non-movable parts – "Smart meters".

Smart meters have a number of advantages as well as disadvantages compared to the traditional mechanical meters. However, the residential (household) meters of this type represent a very small percentage installed in water distribution networks across the world compared to mechanical meters and they can be found only in modern advanced networks, in relatively small numbers or in pilot projects. Smart meters offer a range of logging capabilities, such as air in the system, breaks or leaks in the pipes, minimum and maximum flow, etc.

Such meters are the only meters, which could reliably work under intermittent water supply conditions registering only the water passing through the meter and not the air. However this new technology although suitable for IWS is relatively expensive with an average price for a 15mm diameter domestic meter of the order of 2 to 3 times compared to a corresponding positive displacement meter and to about 3 to 4 times for a corresponding velocity meter depending of course on the quality, characteristics, manufacturer, etc.

8.2 BULK WATER METERS (DMA METERS) TYPES

The supply of large water volumes through transition or trunk mains are normally measured using bulk water meters, whether these are installed at the outlets of water treatment plants, pumping stations, storage reservoirs or on the internal distribution network supplying the DMAs. These bulk meters can either be electromagnetic or ultrasonic or mechanical (Figure 8.4 Woltman type).

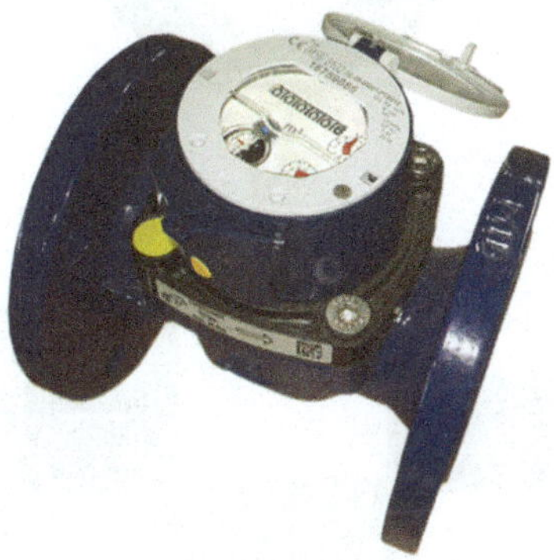

Figure 8.4 Woltman type.

Bulk water meters of the mechanical type are normally installed in water supply networks due to their lower cost compared to other types of meters, electromagnetic, ultrasonic, etc., which is mainly true for water meters lower than 300 mm or less in diameter. However for diameters higher than 300 mm the price difference between the mechanical and other type bulk meters is not so significant.

The presence of air bubbles or turbulent flow, predominant in IWS conditions, may affect the accuracy of measurement of the ultrasonic meters (Figure 8.5). Turbulent flow affects electromagnetic meters (Figure 8.6) as well but to a much lesser extent. Nevertheless, these types of water meters maintain for significant longer period their measuring accuracy and precision, while they are not affected by debris or other transported material which may be present in the water.

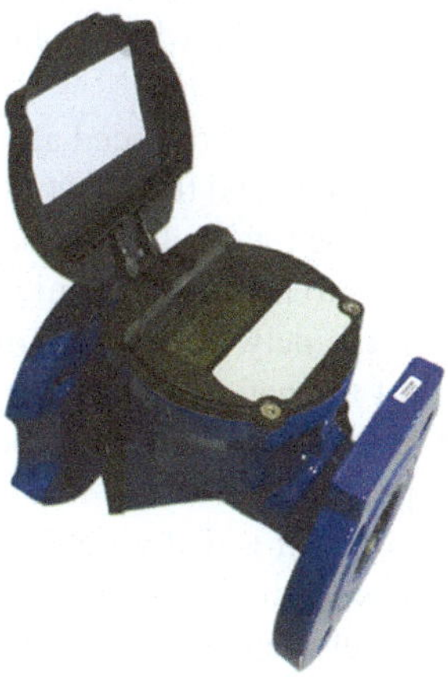

Figure 8.5 Ultrasonic type.

Figure 8.6 Electromagnetic type.

8.3 HOW AIR IN THE NETWORK AFFECTS THE METERS

Under conditions of Intermittent Water Supply a network is depressurized and gradually the pipelines empty, to a large extent, through the continued use of the water by customers as well as through the leaks in the pipes. Simultaneously as water is flowing out of the network the space in the pipelines is filled with air from various locations such as air valves, any holes or cracks in the pipe network as well as from the customer service connections.

The entry of air into the pipes is caused when the network is depressurized, and according to the laws of hydraulics and fluid mechanics, the air will fill the whole volume available inside the pipes to fill the network and to balance pressures.

Accordingly the volume of water in the network is replaced almost entirely by air.

This is reversed when the operation is resumed and the network is again under hydraulic pressure. The water that re-enters the network tends to fill the total volume inside the pipe network and the trapped air is initially compressed and then slowly begins to move as a genuine fluid towards the available exits or the network endpoints in order to escape from the network. Depending on the network design, quantities of air can get "trapped" at the highest parts of the network and to hinder the flow of water on a permanent basis or to create temporary "air pockets", which under the pressure increase above a critical threshold, will revert to future hydraulic surge "hotspots".

The crucial difference of air compared to water is that water has a defined volume, equal to the volume released into the system each time the supply is on, while air tends to occupy the entire volume available inside the pipes. This distinct difference between the two fluids hinders the air management in the network. Air will fill each and every empty space in the pipe network thus affecting many other operating parameters of the network, flow capacity, pressures, etc., in addition to the measurements.

Therefore during the emptying or charging of a network the air which enters or is expelled from the network develops extremely high velocities, the magnitude of which depend on the characteristics of the network, which on many occasions may be higher than the designed capabilities of the meter's measuring mechanism, thus causing the meter to malfunction or rapidly deteriorate to the extent of being completely damaged.

8.4 OBSERVATIONS UNDER IWS CONDITIONS

8.4.1 Domestic (household) meters

The following have been observed in the case of domestic meters operating under IWS.

8.4.1.1 Velocity meters

When only air passes through a mechanical velocity meter, it is registered as if water is passing through the meter, in other words the meter cannot distinguish between the two fluids, air and water. Air passing through the meter sets in motion the registering mechanism and the meter records as if water is passing through with the unavoidable result of customers being charged for quantities of water, which they do not receive.

The velocity type water meters are extremely susceptible to air in the system and their performance under such conditions is inferior to other types of meters. As explained earlier the operating principle of the velocity meters is based on measuring the speed of the passing fluid (water or air), which sets the registering mechanism in motion. This speed is converted through specially calibrated gearing mechanism in the meter to volume, which appears on the display.

In the case of air passing through a velocity meter, as mentioned earlier there is an increased probability of much higher velocities developing than those in the case of water which could even be beyond the structural strength limits of the meter and therefore the problem is magnified with regard to both the measurements as well as the rapid deterioration of the measuring mechanism.

It is evident that with velocity water meters under IWS conditions the registered volume of water does not represent the actual volume received by the user, in fact it is substantially higher in many cases as air passing through the meter is also registered. In addition the life time of the meter under IWS conditions is drastically reduced due to the accelerated wear and tear of the measuring mechanism.

8.4.1.2 Volumetric meters

The manufacturing technology of volumetric water meters differs from that of the velocity meters. In the case of volumetric meters the measuring mechanism does not consists of an impeller which is inserted in the stream, but of a particular volume chamber which fills up and empties continuously as long as the flow is running through the meter. In this way each 'fill-empty' cycle corresponds to a specific volume of water that is registered and accumulated on counter reading. Thus the volumetric meters measure directly the volume of the water passing through, rather than the speed.

Although the measuring principle between velocity and volumetric water meters is different the effects of air when passing through the meters are almost identical in both cases. The only difference, which is effectively negligible, between velocity and volumetric water meters is that the calibrated chamber of the volumetric meter shows slightly greater inertia in the movement of air in relation to water because of the air being compressed within the measuring chamber and thus at low speeds and air pressures the volumetric water meters do not register the quantities of air passing through. Of course this is not valid when the air through the meter is already compressed and passing at high speed, which is mostly the case in intermittent supply conditions when the network is re-pressurised.

In any case with mechanical meters (velocity or volumetric) almost all of the air which passes through the meter will be registered resulting in higher consumption readings and of course higher customer bills since it is not possible to estimate using a reliable method the actual water consumption and to make the necessary adjustments to the customer bill. Also the flow of air through the meters accelerates their wear and reduces their life-time and consequently the reliability of the measurements produced.

8.4.1.3 Smart meters

In the case of water meters with no moving parts the presence of air in the network does not affect the accurate measurement of water in the sense of over registration due to air passing through since these technologies require the internal chamber of the meter to be filled up completely before it starts measuring.

The measuring technology of these water meters is based on electronic measurement and not on mechanical means. Electronic water meters have no moving parts, which are inserted in the water flow. The measurement is achieved with the ultrasonic technology or electromagnetic field or even the residual magnetic field. Consequently these types do not present problems of over-registration due to air passing through nor experience any rapid deterioration of their measurement mechanism.

However, under certain conditions the measurement could be slightly influenced by the timing of the network decompression periods when amounts of air is mixed with the water flow through the measuring instruments. During these conditions, the meters could under-register, or they could register unreasonably high or low values with repeated rapid variations or they could even register no quantities at all, even though some small quantities of water is flowing through the meter.

Thus in this category of meters, air can cause some incorrect measurements but of course on a negligible scale compared to mechanical meters.

It is worth mentioning that very recently advancements in smart meter technology have resulted in the manufacture of a domestic (household) meter which could accurately distinguish between water and air passing through thus resulting in the measurements of the water to be accurate at all times regardless of the amount of air passing through. However, this technology is not widely used mainly due to its relatively high cost.

8.4.2 Bulk meters

The influence of air on bulk (large) meters is completely identical to that of the domestic (household) meters. The essential difference is that the volume of air through the large bulk water meters is statistically far less compared, due to the network layout, to the volume of air that could go through domestic meters. This happens primarily due to the fact that the size of the main pipes is usually quite large and quantities of water remain within these pipes and the bulk meters are normally located at the head of the network therefore the amount of the air in these pipes is usually less, while the air exit points are sufficiently far away from the bulk meters, resulting to the rapid pressurization of the network at the bulk meter location when the water supply is restored.

8.5 REPERCUSSIONS OF AIR ON METERS

The mechanical meters, whether domestic or bulk, suffer mainly from increased air flow velocities through them, sometimes far beyond their mechanical strength limit. The high air flow speed creates significant mechanical stresses on the moving parts of the meters resulting in premature wear of these components or even to their complete malfunctioning in some cases. On the contrary meters with no moving parts, do not suffer from any mechanical stresses given that there are no parts intervening in the flow of water.

The effects of air on water meters, domestic and bulk, installed in water networks operating under IWS conditions are summarized in Table 8.1 and Table 8.2 respectively.

Table 8.1 Domestic water meters.

Meter Type	Meter Category	Effect of Air	
		Metering Accuracy	Ageing of Meter
Velocity type	Mechanical	High	Rapid
Volumetric type	Mechanical	High	Rapid
Ultrasonic type	Electronic/Smart	Low	Not affected
Electromagnetic type	Electronic/Smart	None	Not affected
Residual Magnetic Field type	Electronic/Smart	None	Not affected

Table 8.2 Bulk water meters.

Meter Type	Meter Category	Effect of Air	
		Metering Accuracy	Ageing of Meter
Woltman type	Mechanical	High	Rapid
Ultrasonic	Electronic/Smart	Slight	Not affected
Electromagnetic	Electronic/Smart	None	Not affected

8.6 FACTORS AFFECTING ACCURATE FLOW MEASUREMENTS

The amount of air that enters a distribution network during an interruption of the water supply is directly related to the following main factors:

- Time duration of the interruption.
- Number of possible air entry points into the network (user connections, breaks in pipes, air valves, etc.).
- Size, material and jointing technology of the pipes.
- Length of the network.
- Design, layout and topography of the network.
- Number and functioning condition of protective devices in the network.
- Number of end users and mode of connection to the distribution network (directly, through customer tanks, etc.)

Considering all the above, it is easy to perceive that the calculation of the amount of air that enters a distribution network, is extremely difficult to calculate and could not be easily captured and resolved in a single mathematical formula that would apply to all networks and would deliver reliable and precise results. It is even more difficult to calculate the amount of air that will eventually flow through the water meters and how it will be distributed to each meter.

Therefore it could safely be stated that the volume of air that eventually will pass through the water meters:

- Is not negligible.
- Affects the accuracy of measurements.
- Reduces the lifespan of the meters.
- Increases customer consumption.
- Results in erroneous conclusions when calculating the water balance.
- Undermines the reliability of the entire measurement system.
- Decreases the level of trust between the utility and the end users.

Chapter 9

Communication with stakeholders

Patricia Bakir

9.1 WHY STAKEHOLDER ENGAGEMENT?

Let's start again with the Sri Lanka case study (see Section 10.6). Customers who had household supply received it around 8 hours/day. The remainder obtained water from private wells, public standpipes and neighbours. An excellent first step on this project was to do a household survey to understand the customer's point of view. The survey of 54 households showed that "… 75% had no idea of plans in place to implement continuous supply, though 94% did express their desire for such a service."

The plan to move to 24 × 7 supply included disconnecting all public standpipes, which were used by wealthy and poor alike because they were free. Customers using these standpipes were completely unaware of the plans, and would presumably be very surprised to learn that they would need to install household plumbing at their own cost to accommodate 24 × 7 supply. As noted in the case study, **"… this risked causing public disquiet about a change in service meant to be beneficial."**

The extensive surveying did provide a good opportunity to establish a dialogue with customers. However, as soon as IWS was implemented there were billing complaints, discolouration of the water from leakage repairs, and customers all use much more water.

In short – customers were not properly prepared for the 24 × 7 supply, which could have been greatly alleviated by establishing a stronger dialogue with them right from the start of the project. This may have included a related program to assist households to establish internal plumbing at their homes, at a minimum, one water point within the home.

What can water service providers do to avoid this kind of situation? The first thing they need to do is remember the "first lesson" from the Ghana case study (Section 10.3) **"… a community will do whatever it needs to secure access to water."** The water provider needs to use this as a basis to establish a full, two-way dialogue with customers – not "providing information", but listening to the customer, understanding their needs and responding honestly and transparently to them about any new programs.

 Dealing with the Complex Interrelation of Intermittent Supply and Water Losses
Bambos Charalambous and Chrysi Laspidou
doi: 10.2166/9781780407074_077

In the case of Sri Lanka noted above, households were surveyed, but it does not appear that there was ongoing dialogue with customers, or all the information about the consequences of 24 × 7 supply were explained. Sri Lanka needed *"greater customer engagement prior to continuous supply implementation. Customers need to understand that with increased water availably and usage comes higher bills. Through carefully managing their usage and conserving water the increased number of customer complaints could have been reduced"*.

The following sections in this chapter will provide some general guidelines and principles on stakeholder engagement for moving to 24 × 7. It will not provide a blueprint for specific projects because every situation is different, and full assessment is needed to develop a clear communication plan.

9.2 BASIC PRINCIPLES OF ENGAGEMENT

Stakeholder engagement requires planning just as much – perhaps even more – than engineering and infrastructure development. It requires the same steps one would do for any other kind of project – assessment, planning, implementation and ongoing monitoring and evaluation. There are numerous references available (OECD, 2015 and Water Impact Guidebook, 2012) on the general topic, but none that specifically address IWS or a transition from IWS to 24 × 7. There are a few basic principles required for water service providers to develop a stakeholder engagement plan when transitioning to 24 × 7.

Use communications and engagement specialists – Staff at all levels, including technical staff, can and should be involved in stakeholder engagement activities. However, engineers without specialist experience in communication need the guidance and support of communication and engagement specialists. These can be specialists hired for the duration of a project, but it is better if they are involved in ongoing water provider activities – or a combination of the two.

Assessment – Each country, each situation is different. Do not make assumptions, and do not let staff make assumptions by saying "I know my customers", "I know what the government thinks." Do your homework. Start with an open mind – if you have resources hire an independent agency to do a survey. Use your field staff – they often have a much stronger knowledge of what customers think and will accept than higher-level staff. Identify all key stakeholders (see below for examples) and their potential to be allies or detractors of your project. This will require collecting quantitative (survey) information and qualitative (focus groups, meetings) information to determine what are the concerns of stakeholders and how they can be alleviated through specific activities, through communicating better and through managing the technical transition to 24 × 7 better.

Establishing trust is your main goal – This is key to all stakeholder engagement activities – whether for 24 × 7 or anything else. If the customers and other stakeholders do not trust the water utility, they won't believe anything you tell them. This means they will not easily accept any changes. If they do not have trust in you as a water provider before you start the transition to 24 × 7, there is no reason to expect it will come now.

Focus on two-way dialogue for real engagement – Communication and engagement is not just "posters, brochures, meetings" or only "conveying information". Real communications and engagement includes both receiving information from your stakeholders, and fully engaging them in the decision making process throughout the project cycle.

Do not fear sharing bad news – It is better to share bad news as soon as you have it than to hide it. Stakeholders have their ways of finding information. And in the absence of information, they will make up rumours about what is going to happen. So communicate early and often, good news and bad.

Do not make false promises – Never promise anything you cannot deliver. In the Ghana case study (see Section 10.3) customers had 24 × 7 supply for a short time, then it went back to IWS. This provides a much worse scenario for the water provider than never having 24 × 7 at all.

Focus on target behaviours, not "education and awareness". Awareness is simply knowing about something – it doesn't mean your stakeholders are going to do anything different. What do you want your stakeholders to do? You want the Mayor to support your program. You want customers to pay their bills and conserve water. What are the specific conservation measures you want them to take? What do they need to do in order to take them? You are engaging stakeholders as supporters, and many of them you want to take specific action. Look toward behaviour change and social marketing programs for examples of how to get very specific changes such as water conservation, bill paying, etc.

Identify champions – While conducting the assessment and during technical planning meetings as well as communications planning, identify champions for the new program who have strong influence and can help spread the message to your various target audiences.

Use a variety of communication methods – In addition to face-to-face contacts–which are extremely important – there is a huge range of print and online communication methods for establishing dialogue with stakeholders. Some are more interactive than others. This includes social media such as Facebook, Twitter, websites, Instagram, online news sites, as well as the traditional print, TV and radio media. Advertising, street banners are also used to convey messages, as well as specific events. Specialist assistance will be needed, and different target audiences need different media to be reached. The simplest methods–such as sending mass SMS messages–can also be very effective. SMS messaging–which does not need recipient to have smartphones–can also be used for doing surveys of various stakeholders. You do not need to use all methods. You need to use a variety. Do not by any means avoid face-to-face, personal interaction–this is often the most effective means, especially in relatively small cities and communities. Identify which communication and engagement methods will be most effective for which audiences.

Monitoring and evaluation – Constantly monitor your messages, your engagement strategies, to understand how you as a service agency are perceived, how the messages are understood, whether people are supportive or not, whether they are making the behavior changes needed or not. Establish your method of monitoring and evaluation right from the start so you have a good baseline to understand what's needed as you move along.

Document the process – Take the time to put in writing – even briefly – what you did and how you did it so that others can learn, and so that you can learn for the next activity. Engaging stakeholders for 24 × 7 is a process just like any other stakeholder engagement process, so lessons will be relevant to other situations. For example, the case study from Brazil (see Section 10.1) indicates that customer complaints were not as high as expected because "… most customers understood and supported initiatives of emergency water supply plan and became more conscious about rational use." The case study doesn't tell us why they are more conscious, which would be very valuable information for others who will be implementing programs. Often there are general statements about "participation, engagement, meetings" and yet communications specialists are always on the lookout for specific examples and even materials…

9.3 STAKEHOLDERS

9.3.1 Stakeholder analysis

When planning the transition to 24 × 7 an assessment must include a full stakeholder analysis to understand these issues and how to address them. The term stakeholder refers to anyone, or any group of people, who has an influence and/or interest in transitioning to 24 × 7. This can be people who make the decisions, people who are recipients of services, people affected by construction (who may not be customers), the water provider staff themselves who will have to do their business differently, and many more.

A wide range of stakeholder analysis tools is available through online sources, which can help to identify how to prioritize stakeholder engagement and even means of engagement. The Water Impact Guidebook (see below) has brief chapters on Public Involvement, which include references, and the OECD has produced a useful document outlining the process, including some practical checklists (WRC, 1999;

Charalambous, 2016). Each situation will be very different, each category of stakeholder, and sub-category will need to be reviewed as a project is being developed – not after it has started!

A very simple approach is to categorize stakeholders in one of the following categories:

- High influence, highly supportive
- High influence, against 24 × 7
- Low influence, highly supportive
- Low influence, against 24 × 7

Once this is done, it's easier to plan your stakeholder engagement strategy with highest priority on the ones with high influence – even those against the program who may be able to sabotage it.

When conducting stakeholder analysis, take into account gender considerations and effects on particularly vulnerable populations such as the extreme poor, female headed households and other disadvantaged groups. This will not only ensure that the project cares for these groups, but also will help in planning any communications and engagement interventions. For example, what is the role of men and women in household water management? If women manage the water in the home, but men are the ones who buy the water saving devices, how do we try to get households to buy water saving devices?

9.3.2 Stakeholders

Chapter 6 has outlined some of the main issues and concerns about the transition to 24 × 7. In this section, some of the main categories of stakeholders and some of the issues, which might be of concern to them are described.

All stakeholder groups can be further segmented, as noted below with customers. It is always risky to say "government officials" or "customers" and refer to their attitudes and behaviours as similar to each other. Detailed analysis will help to understand these nuances and further segment stakeholder groups. The main groups only are described below, and are to some extent general because each country and area will have different actors involved.

Decision makers – These vary with each situation, with some of the main decision makers outlined below. Transition to 24 × 7 is going to cost money, disrupt the way of doing business, and cause concern to some customers and other stakeholders. Even making the final decision to go ahead with the project will often include intense lobbying by politicians, government officials, and others. Some of the main decision makers are listed here along with their areas of concern.

- **The water service provider –** In many countries, the government directly manages water supply; there is no private supplier. Whether supply is managed publicly, privately, or by a "publicly owned company", they are both a project implementer and a key stakeholder in the process. The water service provider, as the implementer, needs to take into account its own staff and how their roles might change, what may be required of them, in a transition to 24 × 7. If the internal communication on this process is not transparent and clearly stated, then any external stakeholder engagement process will be much less effective. Also, the water service provider staff, especially field staff, will often have useful input to provide in the planning process; and yet, they are often ignored by senior management or by donors who are controlling the process.
- **Boards** – if private, the water utility board – possibly along with the regulator – will make the final decision on investment in 24 × 7. Boards can be mostly government – in the case of government owned utilities. They can have a strong private sector and/or civil society influence. Each Board member represents a certain group of people/organizations and will come to the decision from that point of view. They will be concerned about rate increases, about whether the utility will be able

to manage 24 × 7 in order to keep their credibility. Boards can be risk-taking, or they can be very risk-averse.

- **Regulators** – In countries with a water regulator needs to balance customer needs and utility needs. The customer needs water 24 × 7, at the lowest possible price. The utility needs to cover its costs. The regulator should be engaged every step of the way in planning for 24 × 7 in order to be a full ally. If the regulator does not support 24 × 7, it won't happen.
- **Other government agencies** – There are usually several government agencies, which play a role in water management, and in citizen satisfaction. The Ministry of Health should provide water quality surveillance – and in the Often there is not a formal "regulator", but there are regulatory agencies such as the Ministry of Health and Environment, who provide water quality surveillance. Ministries of Public Works and Housing, Agriculture, Municipalities/Local Governance often play a role. In countries with restrictive political environments, often there is a need to also keep certain authorities informed of any stakeholder engagement activities. At the start of a project, it is important to map all of these agencies and their specific roles.
- **Local level authorities** – At local level, there are community and municipal authorities who can make or break a program, and often have more influence than national authorities. The water service provider needs to bring these people on board with the project as early as possible, to coordinate both infrastructure works and the planning and adoption process for 24 × 7. These authorities are usually most concerned about pleasing their constituents, so if there will be a huge cost involved, they can often be against any move to 24 × 7.
- **Parliamentarians** – Political figures such as Parliamentarians can also have a very strong influence, and yet often do not have full information in which to make decisions. They will promote great change, but not be in a position to allocate funding; but they can also be an excellent ally if included in the process from the beginning.
- **Civil society** – This is a broad term referring to non-government organisations, influential individuals who may not have official affiliation, and informal community decision-making mechanisms. Understanding the various local level influencers giving them a role to play in the promotion of 24 × 7 supply is often the best way to also ensure their long-term support. They are also an invaluable source of information on customer attitudes and current practices, and should be consulted regularly.

Customers – There is a wide variety of customers, usually divided in very general terms into: commercial, industrial and household customers. Even within these categories, there can be great variation of their needs, their current supply, and how they will perceive and adapt to a change to 24 × 7. There will always be a need to further segment the main categories below depending on socio-economic and often also political factors.

- **Industry** – Industry is treated as a separate category due to their particular needs in terms of water supply and wastewater needs. Industries often run their factories 24 hours/day and use much more water than other customers. Often, industries have their own, separate supplies; but if they are connected to the municipal supply, then their particular needs must be taken into account if there will be any major change. They also have very specific wastewater needs, due to the nature of their discharges, and these can change if more water is supplied under a 24 × 7 regime. Small industries will have different needs from large industries, and often will be connected to municipal supply. Large industry owners often have a strong political influence – and some of them are politicians.
- ***Commercial customers*** – This refers to a wide range of business customers, which can vary greatly in size: shops/malls; restaurants; small/large hotels; small/large office buildings, and many others. As with industry, the large ones may have their own water supplies, but many will not. Also, some commercial customers will be those who, legally or illegally, sell water on to other customers – they

will have a big stake in not supporting a 24 × 7 supply. As with industry, these customers can be part of the wealthy and political elite in a country or region.

- **Households –** Households are usually segmented by water service providers by their water use patterns. It is assumed that the lower use households are the lower income households. However, this is not always the case, since this can also be a reflection of house size, how much households supplement their water from other sources and other factors. A description of the various topics of concern for households is presented in Section 2.1.3 above. Households need to be segmented according to how much they use of municipal supply, how much they use per capita from all supplies (to identify at risk households), and how much water they use for other uses such as kitchen gardens/irrigation and landscaping. There also may be socio-political factors, if different areas of a service area have different ethnic groups. Also, there are often illegal settlements, which get some public water supply. Each situation is very different – the important principle here is to not just categorize household customers as all the same. The Water Impact Guidebook (Charalambous, 2016) specifically addresses Water for the Poor, an important issue to take into account when transitioning to 24 × 7.

The water service provider – Often it's the utility itself which is against 24 × 7 supply. They worry there will be increased leakage and therefore more water lost with 24 × 7 supply. They fear lower revenues due to less water sales; yet at the same time they are supposed to be encouraging customers to conserve. Change is always difficult in large organizations – even "good change" – such as more frequent supply. The utility needs its own internal communication plan for its staff, supported by the regulator and Board, to (1) gain support for 24 × 7, and (2) adopt the necessary internal management changes to fully implement 24 × 7. Also, an extremely important utility staff group is the field staff who has regular contact with customers. They are often neglected and not considered important – but they have a huge role to play in both implementation and in communicating with customers. If a customer asks "I hear 24 × 7 supply is coming" and the meter reader says "Well, that's the first I heard of it. I'll believe it when I see it," then all the media work will be useless.

Private sector – the private sector does not just include industrial and business customers. It also includes private water vendors who may see a loss of business along with the new 24 × 7 supply. Their level of influence and commitment to 24 × 7 needs to be assessed, since water vendors are quite powerful in some countries and communities. If they are against 24 × 7, ways to alleviate their objections need to be found, which could include helping them identify alternate business opportunities.

Media – Print, TV and radio media are still very influential in many countries, despite the additional role of social media sites. It is never a good idea to go around media and avoid them, as they can do a lot of harm to your programs. Sending press releases never hurts – but the water provider should, long before a 24 × 7 program, have established a good relationship with key media influencers. This is especially important when one will be trying to persuade other stakeholders – such as politicians – of the importance of increased infrastructure investment. Politicians often have a stronger relationship with the media and directly and indirectly affect customers in that manner – so it's important to have an equally strong relationship with them. Don't underestimate the power of social media – one Tweet, one Facebook post negative about the project can go viral and do great damage. Establishing a strong social media presence and monitoring of all media types is important.

9. 4 REFERENCES

OECD (2015). Stakeholder Engagement for Inclusive Water Governance, OECD Studies on Water, OECD Publishing, Paris. http://www.medspring.eu/sites/default/files/Stakeholder-engagement-for-inclusive-water-governance.pdf; http://dx.doi.org/10.1787/9789264231122-en.

The Water Impact Guidebook (2012). Capacity Development to Enhance Commercial and Technical Management in Water and Wastewater Utilities. GIZ 2012, Germany, ISBN: 978-3-8236-1649-8.

Chapter 10

Case studies

Francisco J. Paracampos, Debora Souares and Natally Annunciato Siqueira

10.1 BRAZIL: INTERMITTENT WATER SUPPLY – THE REAL BILL COMES LATER

10.1.1 Abstract

The case study describes a situation faced by the Metropolitan Region of São Paulo, an extremely dry two-year period (2014–2015). An event such as this had a probability as low as once in two hundred and fifty years to occur. The reasonably robust water supply system of São Paulo was under pressure for delivering water to 20 million inhabitants. The main aim was to keep continuous supply for twenty four hours, since the damaging effects of rotational supply (intermittent supply) were well known to the water utility, but as the crisis persisted for a long period, and the predictions about the following hydrological year were consistently discouraging, radical decisions had to be taken.

This case study covers the strategy and effect of the operational regimes adopted during the drought on the burst frequencies in the distribution network, during the following three marked phases, namely: normal operation, drought operation and recovering operation.

10.1.2 Scenario

The water supply system in the Metropolitan Region of São Paulo (MRSP) consists of nine integrated water source systems: Alto Cotia, Baixo Cotia, Alto Tietê, Guarapiranga, Rio Claro, Rio Grande, Ribeirão da Estiva, Embú-Guaçú e Cantareira which are used to produce and deliver water for 20 million inhabitants to 35 different cities.

Cantareira is the main water source system, which has production capacity of 33m³/s, supplying about 47% of MRSP and about 65% of São Paulo city.

Bambos Charalambous and Chrysi Laspidou
doi: 10.2166/9781780407074_083

During the hydrological period from October 2013 to March 2014 an unusual dry weather was experienced, which statistically had very low probability to occur. This situation persisted throughout the whole year of 2014 and became even worst in 2015.

To understand the magnitude of the challenge for keeping the water supply system on line, the Brazilian National Agency of Water which has the responsibility of managing and regulating the water resources as well as promoting sustainable water use, imposed gradual reduction in water withdrawals from Cantareira water source reaching the amount of 13.5 m^3/s in March 2015 (Figure 10.1). This value was 41% lower than the system potential. Additionally, the second largest water source, Alto Tietê (13 m^3/s), was also impaired by the adverse climate, and the production was reduced by 30%. The third largest, Guarapiranga water source was in a better condition, and was upgraded by 2.5 m^3/s in order to minimize the reduction in production from the other two water sources. This meant that Guarapiranga water source was over exploited during this period although this was a very small positive contribution to the magnitude of the entire problem.

Figure 10.1 Cantareira reservoir during the drought period.

The area affected in this drought scenario, was the central area of São Paulo, subject of this case study, which was highly impacted by the restricted availability of water resources since Cantareira is the main water source for this area of São Paulo. As restrictions in water catchment were necessary, actions to mitigate the water shortage problem had to be taken.

10.1.3 Emergency water supply plan

As it is known, a rotational supply in a given water distribution network may be a solution for lack of source water since consumption would decrease as a result. However its damaging impact alongside the health risks, the delay in water reaching remote areas of the distribution network and the increase in pipe bursts frequency influenced the decision-making process. In this case of the central area of São Paulo, the plan included four main actions:

- Consumer motivation to reduce water consumption through a bonus program, applying discount on the bill;
- Reduction of the area supplied by Cantareira system and expansion of neighbouring systems;
- Intensification in water loss reduction activities;
- Implementation of a Pressure Reduction Program (PRP).

The PRP was a daily pressure head decrease in the distribution network prioritizing periods with lower demand, however keeping the pipeline full with pressure head just above zero thus avoiding emptying of the network.

Since supply has been re-established at the end of the crisis in February 2016, it was necessary to investigate the effects of PRP. Thus, this case study will focus on the lessons learned and results obtained in the Central Business Unit of SABESP, exploring how the strategy and effects of the operation rules adopted has impacted the distribution network. The analyses covers three marked, phases: normal operation, drought operation and recovery operation phase evaluating the bursts and leak frequency.

10.1.4 Water supply characteristics

The case study analyses the area covered by the Central Business Unit of SABESP, which comprises São Paulo downtown area and the expanded part of the Eastern city of São Paulo, where there is a very dense urbanization with 122 connections per km of mains and a great social-economic diversity. It has a land area of 281 km^2 and a population of about 3 million inhabitants. However for covering much of the consolidated urban sprawl, its main feature is the floating population of 1.3 million people per day.

The Central Business Unit of SABESP has the following characteristics:

- 759,859 active connections comprising mainly high density polyethylene
- 6,220 km of distribution mains comprising mainly cast iron pipe
- 9 booster stations
- 26 bulk supply zones
- 213 Pressure Regulating Valves (PRVs) covering around 41% of the distribution network

The infrastructure in this part of the city is the oldest in the MRSP. About 62% of distribution mains are more than 40 years old thus making real losses the predominant problem in Central Business Unit.

The average pressure head of the system is 39 m, but significant pressure drops occur due to high head losses during the peak hours in a substantial portion of the water distribution network.

The typical domestic plumbing arrangement, as defined by the Brazilian Association of Technical Guidance, has consumer roof tanks, which accumulate a volume of water related to the daily average consumption of the household.

10.1.5 Pressure reduction program implementation

The objective of the Pressure Reduction Program (PRP) was to reduce the water flow demand in the system but also to control the negative impacts from the program with assertive actions.

The major source of problems and complaints was due to the delay in recharging the water mains. For mitigating this problem the pipes were kept full thus avoiding their continuous emptying and refilling.

From the customers stand point, it was strongly observed to respond to any complaints with actions such as field inspections, surveys, logging of the pressure head in order to provide detailed feedback to the customers.

Before applying pressure reduction in each bulk zone, the first step was to test a small pressure reduction in some zones. The responses of the water supply distribution network were unknown and rather different from one zone to another. With the initial feedback evaluation, as well as analyses of the pressure head behavior at critical points, the pressure head reduction was further increased which was the final step in the PRP implementation Another important point of the program was the refinement work, both operational and functional, carried out at the pressure reduction valves (PRV).

The program was fully implemented in March 2014. At this stage, the pressure reduction period had an average duration of eight hours per day applied primarily at night, when the demand was lower. As the drought problem worsened, the average reduction period had to be extended to fourteen hours per day starting in February 2015 up to February 2016 when the water sources reservoirs had in storage sufficient volumes. Since February 2016, knowing the benefits of pressure management in water loss reduction, the program has been maintained with an average reduction of six hours per day.

It is well known that pressure management is extremely important in controlling burst frequencies on mains and services. Higher pressures and pressure transients can increase this frequency.

In response of PRP, the average pressure head in Central Business Area dropped to 31 m in 2015, and at (2016) was at 33 m. However the difference between maximum and minimum pressure heads had increased from 11 m in 2013 to 26 m in 2015.

Another important result of the program is the difference in the system input volume (SIV) in study area, which decreased from 14.93 m³/s, before the water crisis period, to 11.15 m³/s, during the worst phase of the crisis (Figure 10.2).

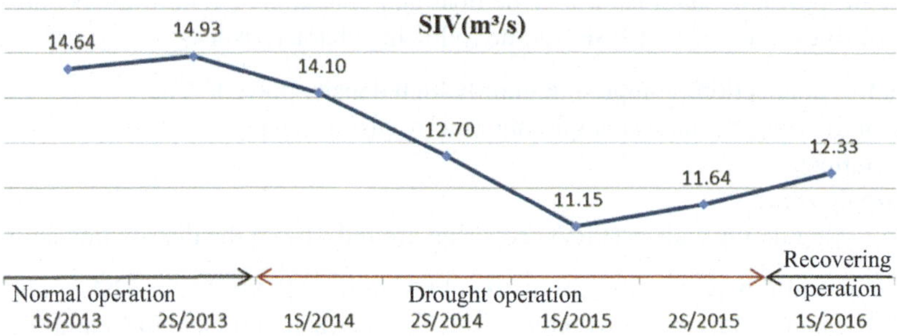

Figure 10.2 Semiannual SIV in study area.

10.1.6 Data analysis of reported and unreported leaks

Reported and unreported leak records of the Central Business Unit for the last ten year period (2007–2016) are presented in Table 10.1. The number of leaks on mains per 100 kilometers per year varies from 69.41 to 91.01, five to seven times worse than the recommended value by the Water Loss Specialist Group (WLSG) of International Water Association (IWA) of 13 leaks per 100 kilometers per year. For service connections, the data in Table 10.1 shows that the actual number of leaks varies from 23.50 to 48.82, eight to sixteen times higher than recommended value by the IWA WLSG of 3 leaks per 1,000 connections per year (Lambert *et al.* 1999).

The impact of pipe burst and leak frequency during three marked, phases: normal operation (year 2013), drought operation (years 2014 & 2015) and recovering operation phase (year 2016) was analyzed. For each one of the above three phases, data for the first six months of each year were used in order to evaluate similar conditions since recovering operation phase is in progress.

Analyzing the burst frequency for the Central Business Unit area, a high frequency decrease from 2013 to 2014, when the program started, is evident as shown in Table 10.2, and graphically in Figure 10.3. In 2015, when the PRP period was longer, the frequencies began to increase. At this phase the difference between maximum and minimum pressure heads was more than double indicating that transient pressures in the network might be a possible cause of leaks.

Table 10.1 Reported and unreported leaks in mains and service connections of the central business unit.

Year	Number of Leaks in Mains	Length of Mains (km)	Number of Leaks/ 100 km/year	Number of Leaks in Service Connections	Total Number of Service Connections	Number of Leaks/1,000 Service Connections
2007	4,903	5,803	84.49	34,015	696,697	48.82
2008	4,346	5,849	74.31	28,392	705,396	40.25
2009	4,257	5,902	72.13	24,566	710,765	34.56
2010	4,135	5,957	69.41	22,123	719,803	30.73
2011	5,063	5,981	84.66	22,906	730,98	31.34
2012	5,334	5,861	91.01	23,211	738,435	31.43
2013	5,134	5,884	87.25	19,858	745,747	26.63
2014	4,803	5,905	81.34	21,178	753,094	28.12
2015	4,735	5,897	80.30	17,820	758,334	23.50
2016*	4,698	6,220	75.53	19,582	759,859	25.77

*Data extrapolated from semester to a year.

Table 10.2 Leak frequency in mains – central business unit (number per 100 km per year).

	Normal Operation (2013)	Drought Operation (2014)	Drought Operation (2015)	Recovery Operation (2016)	Network Length (km)
Central Business Unit	93.9	70.9	78.1	75.5	6,220
DMA with PRVs	107.0	83.5	86.9	82.2	2,576
DMA without PRVs	84.6	62.0	71.9	70.8	3,644

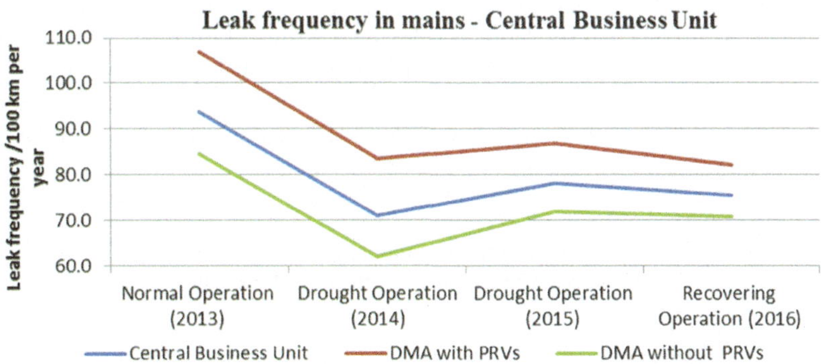

Figure 10.3 Leak frequency in mains – Central Business Unit (Number per 100 km per year).

To better analyse the leak frequencies in critical sectors, namely those which have 70% of the network with pipes older than 40 years, with total length of 2.708 km, a separate study was carried out (Table 10.3 and Figure 10.4). It is evident that in areas covered with pressure reduction valves (PRVs) the effects of PRP are mitigated. In contrast, the leak frequency in areas not covered by PRVs is higher and exceeds burst characteristics of the normal operation.

Table 10.3 Leak frequency in mains – critical sectors (number per 100 km per year).

	Normal Operation 2013	Drought Operation 2014	Drought Operation 2015	Recovery Operation 2016	Network Length (km)
Critical Sectors	86.6	74.6	84.7	87.2	2,708
DMA with PRVs	101.6	97.6	92.0	90.2	892
DMA without PRVs	79.2	63.3	81.2	85.8	1,816

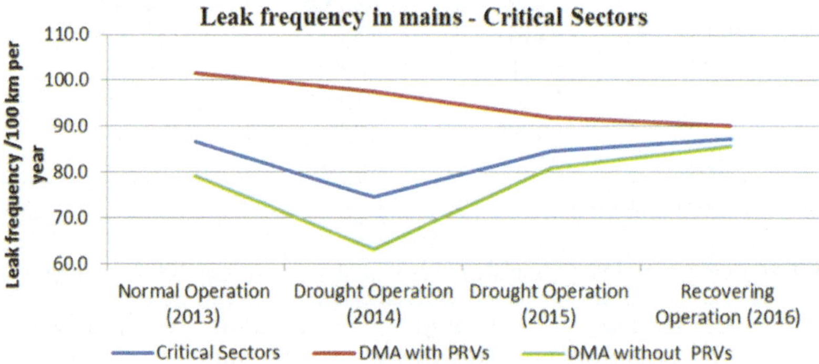

Figure 10.4 Leak frequency in mains – Critical Sectors (Number per 100 km per year).

Furthermore an unexpected observation was made since the areas protected by a PRV have higher number of bursts than those without this protection. A probable cause of this may be that these areas might have a deteriorated infrastructure and the installation of PRV occurred at a much later stage, and then full response was not achieved.

Also, these DMAs with PRV had operational rules with strong closures of flow and pressures daily. Consequently an additional effort was applied over such parts of distribution network.

Comparing the leaks in mains of the entire Central Business Unit versus critical sectors zones, after the drought operation (2014–2016), it can be seen that higher leak incidences occurred in the Critical Sectors where the network is more than 40 years old (Figure 10.5).

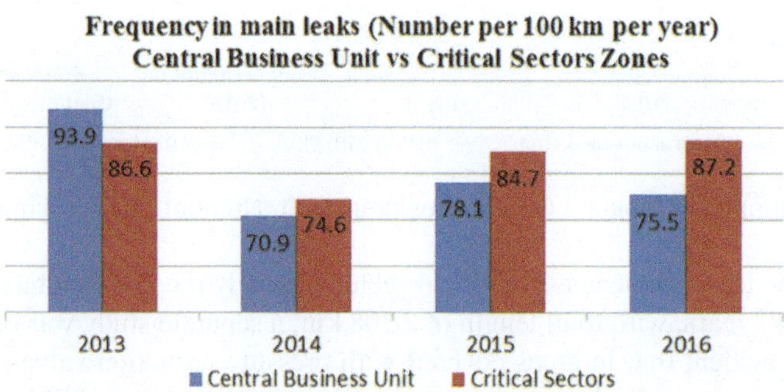

Figure 10.5 Comparison of number of leaks between central business unit versus critical sectors zones.

Similar analysis was conducted for service connections. Table 10.4 and Figure 10.6 show leak frequencies in service connections of the Central Business Unit. In drought operation (2014–2015) the number of leaks show a decrease with a significant reduction in 2015. In the recovering operation phase (2016), the number of leaks has increased particularly in zones not covered by PRVs.

Table 10.4 Leak frequency in service connections – central business unit (number per 1,000 connections per year).

	Normal Operation 2013	Drought Operation 2014	Drought Operation 2015	Recovery Operation 2016	Number of Service Connections
Central Business Unit	29.79	28.25	23.70	25.77	759,859
DMA with PRVs	31.23	27.85	24.74	24.32	281,594
DMA without PRVs	28.94	28.48	23.08	26.62	478,265

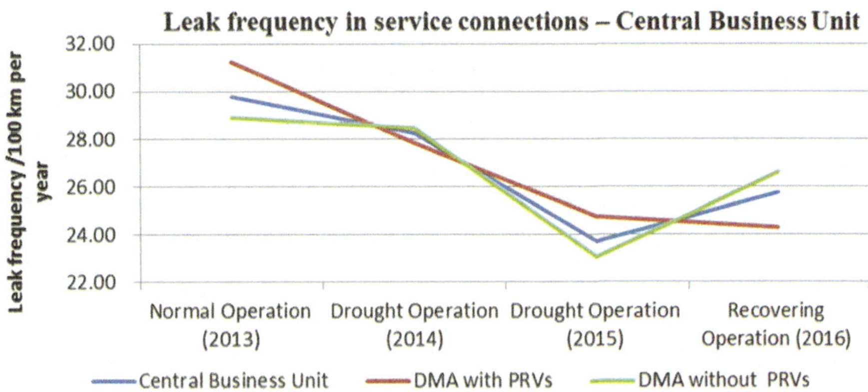

Figure 10.6 Leak frequency in service connections – Central Business (Number per 100 km per year).

Carrying out similar evaluation in critical zones (Table 10.5 and Figure 10.7), year 2014 presents a small increase in the number of leaks, whereas year 2015 shows a noticeable decrease and in year 2016 it reverts to an increase. Again the importance of PRVs in maintaining stable conditions is shown in year 2015 and 2016.

Table 10.5 Leak frequency in service connections – Critical Sectors (Number per 1,000 connections per year).

	Normal Operation 2013	Drought Operation 2014	Drought Operation 2015	Recovering Operation 2016	Number of Service Connections
Critical Sectors	29.67	30.91	25.51	27.79	332,981
DMC with PRVs	30.65	32.31	26.70	24.99	103,818
DMC without PRVs	29.23	30.28	24.97	29.06	229,163

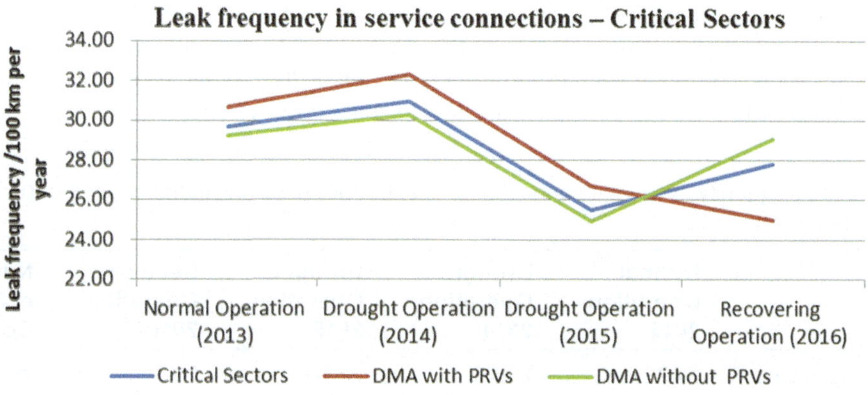

Figure 10.7 Leak frequency in service connections – Critical Sectors (Number per 100 km per year).

10.1.7 Effects of operational rules

The main concept of applying a severe reduction in the pressure head, for a limited period during the day as explained above, proved to be efficient enough in coping with the extreme limitation in system input volume and caused much less upheaval to customers compared to a traditional rotational supply. The volume of water per household, in accordance with the Brazilian Standards for consumer tanks, could support 24 hours household consumption without been fed by the public water system.

The full implementation of the concept was based on the interconnectivity of the distribution network. However a number of unknown limitations appeared as the system was stressed and a fast response had to be carried out. For instance, lack of pressure due to undersized pipes, lack of air valves, as well as regular disruption in any water distribution system for fixing leaks, which had to be fixed in twelve or fourteen hours, whilst the network was full.

The customers' complaints were higher, mainly during the initial phase, about five times those in a regular season, but much less than expected.

The long-term effect will be seen later on, as projected from the previous chapter. For sure, the stress over the distribution system will provoke even higher frequency of failures that can already be observed.

10.1.8 Conclusions

Given an unpredictable drought, only a few actions to mitigate its effects could have been taken in a short time. At the end of the day a compromise between availability of water resource and limitation on customer demand was achieved.

The management of the pressure head in the network provided unexpected positive results in a short term (2 years) period, however the life time expectancy of the pipes was strongly affect mostly in the critical sectors of the network, and the adverse effects can already be seen by the increased number of leaks occurring in the recovering operation phase (year 2016).

Most customers understood and supported the activities of emergency water supply plan, and became aware of the importance of rational water use.

To minimize the impact of drought on water availability, the company is investing in new water sources as well as in works facilitating water transfer amongst the various sources. Additionally, the water loss control program continues as the main long term action for improved efficiency.

10.2 CYPRUS: THE EFFECTS OF INTERMITTENT SUPPLY ON WATER DISTRIBUTION NETWORKS

Bambos Charalambous

10.2.1 Abstract

The world's population is increasing at a tremendous rate, the world's renewable water resources are reducing rapidly, the gap between supply and demand is widening with urbanisation and climate change making it even wider. This case study reviews how the Water Board of Lemesos, Cyprus, a water utility with a proven history of managing extremely well its distribution network, was forced due to water shortage conditions to have intermittent supply, providing water 3 times a week for about 12 hours each time. Although intermittent supply is perhaps the last measure to be taken in conditions of water shortage, it is however a situation worthwhile avoiding through proactive planning and timely response to critical conditions. The adverse effects of intermittent supply on customer service and on the integrity of the distribution network, increase in number of bursts and leakage, as well as the financial repercussions to the utility are also discussed based on the experiences gained from the intermittent supply measures taken during the water shortage periods faced by the Water Board in the last twenty years.

Keywords: Water shortage; Intermittent supply; Leakage

10.2.2 Introduction

A drought can be defined as a prolonged period of unusually dry weather in an area where some rain might normally be expected. Droughts involve inter alia, water shortages, crop damage, stream flow reduction and depletion of groundwater and soil moisture. A drought happens when a period of low rainfall leads to a shortage of water. It is starting when total rainfall is well below average for several months. A balance must be maintained between the water taken out for supply and that being replaced by surface run-off. Normally the surface run-off during the winter far exceeds demand for supply, so that the excess water can be stored and used when surface run-off is less than demand from consumers, normally during the summer period.

Usually water shortage is declared when the water supplies fall short of meeting water demands. This situation usually prevails in arid and semi-arid areas of our planet where precipitation has been steadily declining or where management of limited water resources has been wasteful and unwise.

Intermittent water supply may be defined as a piped water supply service that delivers water to users for less than 24 hours in one day. It is a type of service which is very common in LAMI countries but little found in the rest of the world. In an intermittent supply situation the consumers secure their water supply through the use of ground or roof tanks, where water is stored during the length of time that the supply is provided.

It is worth noting that intermittent water supply is enforced not only in cases where there is water shortage but also where the hydraulic capacity of a network is such that cannot satisfy demand as well as in cases where the network is severely deteriorated.

In many instances there is no indication how long intermittent supply measures will be in place. The hydrological conditions in each case could impact adversely on water supply for years in which case conserving as much as possible the limited water resources may not be the long term solutions but it may be necessary to add to the water balance new non-conventional water resources. In many countries water shortage problems were overcome through the desalination of brackish or saline water. Of course exploring every potential water source available may be the only solution in many instances, but conservation and leakage reduction is always one of the least expensive and quickest solutions to ensuring that water will be available when needed.

10.2.3 Water loss minimisation

Reducing losses from distribution networks is of the utmost importance and water utilities must recognise this and respond positively. Efficient and effective water loss control should be recognised as a first priority for improving potable water supply.

Accounting for water is the first step that must be made by any utility. It is imperative that an accurate and comprehensive metering system is in place for registering all water along the chain from production to the consumer, including measurement of the water produced and/or imported, water flow in and out of treatment plants and storage reservoirs and into the zones and district metered areas. It is imperative to eliminate or minimise Authorised Un-Metered Consumption thus achieving the highest possible accuracy in accounting for all water produced. Apparent and Real losses must be analysed and action taken as necessary to reduce the Apparent Loss component thus increasing the utility revenues as well as applying the most cost effective leakage programme which reduces leakage to an economically, environmentally and socially acceptable level.

Decision makers at all levels in water utilities must understand that any water loss control strategy, in order to be effective, must be a continuous activity based on a long-term strategy and should form an integral part of the utility's vision. The success of the strategy will inevitably depend on the commitment and dedication at all levels within the utility and of course on the adoption of appropriate strategies and techniques. A successful strategy is one that maintains the distribution network in a proper working order, reducing and maintaining leakage at an economic level, and of course providing the required level of service to all consumers.

10.2.4 The water board of lemesos case study

10.2.4.1 The distribution network

The Water Board of Lemesos is a non-profit, semi-government organisation charged with the responsibility of supplying potable water to the town and environs of Lemesos. The main activities of the Water Board are: planning and execution of technical projects, operation and maintenance of the water production and water supply systems and all associated financial services including collection of water revenues and determination of water tariffs.

The development of the distribution network took place in an organised fashion with new areas of supply being incorporated into their respective pressure zones, strictly governed by the areas ground contours. Each pressure zone is subdivided into DMA's, a total of 60 are currently operating, which have a single metered source with physical discontinuity of pipe work at boundaries.

In late 1980 the Water Board embarked on a detailed programme of leakage management. Since 2002 the Water Board has adopted the practices and methodologies advocated by the IWA Water Loss Group. The efforts made and importance placed by the Water Board for proper leakage management is reflected in the reduction of the non-revenue water over the years and in the improved operational performance of its network. In 2007 the Board had an ILI below 2 and real losses under 100 litres/connection/day.

10.2.4.2 Water supply conditions

The Water Board realised that water conservation is not to be equated with temporary restrictions on customer water use. Although water restrictions can be a useful emergency tool for drought management or water shortage situation, water conservation programs concentrate on continuous improvements in water use efficiency. To this end the Water Board embarked on a promotional campaign through television, radio and leaflets to increase public awareness for water conservation.

In 1991, the government legislated against the use of hosepipe for washing cars and pavements at all times, a law, which the Water Board enforces during drought periods only. During the drought period of 1997–2000 the government was forced to announce in early 1997 a reduction of 20% for potable water and 40% for irrigation water supplies. In 1998 the water situation became worse and the restriction measures became more stringent as the available quantities of water were diminishing and the government, much to the discontent of the public, went ahead with further measures, enforcing greater restrictions to the water supplies with targeted figures of 28% for potable water and 56% for irrigation use.

The Water Board of Lemesos always responded promptly to the government's declared drought measures and imposed each time restrictions to the continuous supply. In February 1997 restricted supply to consumers to four days a week. In 1998 due to limited water reserves the Water Board was forced to decrease further the availability of water reducing the time of the water being available to the consumers to 12 hours every 48 hours. These measures were in force until the end of 2000. These actions resulted in an overall reduction in the use of domestic water of approximately 15% per annum which proved that the domestic demand is near enough inelastic.

It was thought that after the above 4 year period of intermittent supply, measures would have been taken so that there will be sufficiency of water quantities for at least the domestic needs. However, this was not the case and 8 years later, in 2008, water cuts for domestic use were imposed once again. Water was brought with tanker boats from Athens, Greece. For many this dire situation is blamed on the politicians for not taking the right decision at the right time. For others, there is the argument that there has been mismanagement of the water reserves prior to the drought period and not sufficient forethought has gone into the planning of such a severe drought taking place so soon after the last one. In any case the fact remains that the Water Board of Lemesos even though it has improved its network to such an extent that it is consider amongst the world's best with losses from the network being extremely low, it was faced with a situation beyond its control. Of course the situation would have been a lot worse if the Board did not continuously improve its network in order to minimise losses thus saving valuable quantities of water.

Intermittent supply therefore, may have seemed to be the short term answer to the water shortage situations faced in the last 20 years, however in the following sections the adverse effects on the integrity of the distribution network of such actions are discussed and quantified based on the experiences gained.

10.2.4.3 Effects of intermittent supply

Analysis of case study data showed that there was a large increase in the number of reported pipe breaks during the period of intermittent supply. In order to quantify these a comparison was made for a large number of District Metered Areas, 20 in total, between the breaks reported in 2007, before the intermittent supply was applied, and those reported in 2010, the first year immediately after the measures were lifted. The results are shown in Table 10.6 covering both mains and service connections.

Table 10.6 Effect of intermittent supply on reported pipe bursts.

Description	Number of Reported Breaks		
	Before (Year 2007)	After (Year 2010)	% Increase
Mains	14 per 100 km	42 per 100 km	200
Service Connections	15.5 per 1,000	29.7 per 1,000	100

This comparison showed that the number of breaks on mains increased from an average of 14 per 100 km of mains to 42 per 100 km of mains, an increase of 200%. Similarly the number of reported service connection breaks increased from an average of 15.5 per 1,000 connections to an average of 29.7 per 1,000 connections an increase of approximately 100%.

Of course, in addition to the reported breaks in 2010, there is a significant number of breaks, caused by the frequent emptying and refilling of the network, which do not come to the surface since the network is not pressurised for any significant length of time to force the water to come to the surface or to have the opportunity to locate these through active leakage control.

The Minimum Night Flow has increased in all District Metered Areas, a typical example being District Metered Area (DMA) 129. Figure 10.8 shows the Minimum Night Flow before and after intermittent supply for this DMA. From the graph it is evident that leakage in this DMA increased from 10 m^3/hr to 20 m^3/hr.

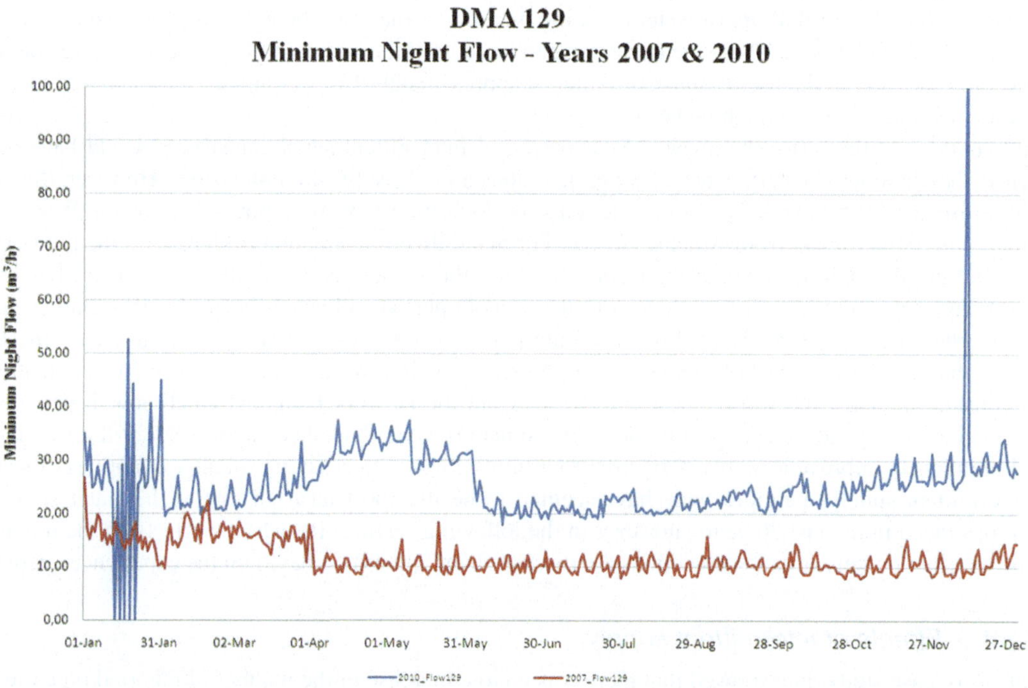

Figure 10.8 Minimum night flow graph for DMA 129.

This situation is typical of all DMAs with relatively old, about 40 years, network. In order to reduce leakage the Water Board applied a program of Active Leakage Control commencing with Districts with the highest increase in the equivalent number of pipe bursts.

Calculations have shown that the overall increase in leakage due to the intermittent supply measures was of the order of 9%. This figure was substantiated in an analysis using a 'top down' and 'bottom up' approach for 2007, before the intermittent supply was applied, and for 2010, the first year immediately after the measures were lifted and 24 × 7 supply was again in place.

Figure 10.9 shows the total Minimum Night Flow before (blue colour) and after (red colour) the intermittent supply. It is obvious that there has been a significant increase in the Minimum Night Flow, which could only be attributed to the additional breaks, which the network suffered during the intermittent supply period.

These non-reported breaks will have to be located through active leakage control activities and repaired in order to reduce the level of leakage to the level prior to the application of intermittent supply measures.

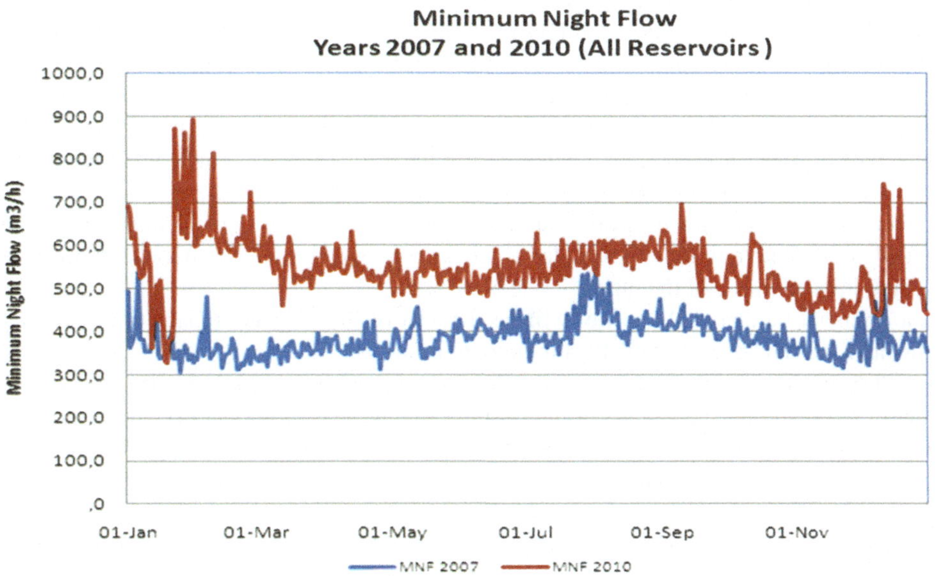

Figure 10.9 Minimum night flow graph for all reservoirs.

Further evidence from the case study to substantiate the increase in leakage due to the intermittent supply measures is given in Table 10.7 which shows an increase of 12.8% in the System Input Volume in Year 2010 after intermittent supply measures were lifted compared to Year 2007 before the intermittent supply measures were applied without a corresponding increase in customer consumption. In fact the customer consumption was slightly less (1.2%) in year 2010 compared to year 2007. It is also evident from Table 10.7 that the System Input Volume in the first year, 2008, of Intermittent supply decreased by 17.5% whereas in the second year, 2009, by 9.1% indicating that the number of breaks in the network increased in number in the second year of intermittent supply. This is substantiated by the fact that the reduction in the customer consumption remained effectively the same for both years of intermittent supply.

Table 10.7 System input volume vs customer consumption.

Year	System Input Volume	Customer Consumption
2007 Before Intermittent Supply	Base line 0%	Base line 0%
2008 Intermittent Supply	−17.5%	−9.2%
2009 Intermittent Supply	−9.1%	−8.9%
2010 After Intermittent Supply	+12.8%	−1.2%

Furthermore numerous complaints were received from disgruntled consumers regarding quality problems and of course lack of pressure during intermittent supply. Needless to say that intermittent supply caused serious disruption and upheaval to the daily activities of people whether these were at home or at work.

10.2.5 Cost of intermittent supply

The implementation of intermittent supply has a direct financial cost to the water utility in addition to the loss of revenue due to the decrease in the sales of water.

The loss of revenue to the utility is a direct result of the conservation of water by the customers. Of course, the corresponding system volume is also less which means that overall financial loss will be the difference between the selling and buying/producing the water. The cost of water, which is lost through the additional leakage caused by the intermittent supply operation depends on the running time of each leak and the cost of water. It is however a major cost to the utility and one which will continue to burden the utility until the additional leaks are found and successfully repaired.

The direct costs include, amongst other additional operational costs for opening and closing sluice valves to implement water rationing the cost for repairing reported breaks caused to the network due to the frequent emptying and filling of the pipes.

A major financial burden to the utility will be the cost of locating and repairing unreported breaks through an intensive and concentrated effort in order to minimise the running time of the additional leaks.

The Board estimated that the overall average loss of revenue due to the reduction in water sales was of the order of €300.000 for the 2 years of intermittent supply. The additional expenses paid in overtime to staff for the same period of the 2 years was estimated at €365.000 whereas the cost of repairing the additional reported breaks was of the order of €325000.

It is also estimated that the volume of water which is lost due to the increase in leakage when continuous supply was established was estimated to be approximately 1.655.000 m^3 per year which is equivalent to €1.325.000.

10.2.6 Conclusions

It is evident from the results presented in this paper that although intermittent water supply may seem to be a solution to a water shortage situation in overall terms the water balance is adversely affected. Supplying less quantity in an intermittent manner causes such deterioration to the network that when continuous supply is re-established additional quantities are lost through increased leakage, which in fact places an added financial burden on the utility.

It is therefore evident that no matter how good a network is, intermittent supply operation has definitely a detrimental effect on its integrity and in addition the amount of water 'saved' is later 'lost' and in greater quantities through increased levels of leakage. Such operational conditions should be avoided especially in pipeline networks that have been designed for continuous supply.

In addition it has been shown that the domestic demand is in effect inelastic and in fact the quantities of water saved by the customers were very small. It is the authors' opinion that perhaps better results could be achieved through a structured conservation programme rather than intermittent supply. Of course such programmes are to be introduced as part of an overall strategy for water conservation both on the supply and demand side.

10.3 GHANA: THE CHALLENGES OF TRANSITIONING TO 24 × 7

Joe Dalton

Keywords: Water losses; Continuous supply; Network efficiency

10.3.1 Introduction

The city of Tamale, the capital of the Northern Region of Ghana, was served with water from an intake on the White Volta River located at Nawuni, 37 kilometres north of the city (Figure 10.10).

Figure 10.10 Raw water intake at Nawuni.

The treatment plant at the nearby village of Dalun supplied 19 million litres of water per day (MLD) to the city via a 450 mm transmission main. Multiple villages that lived along this transmission main, the "Dalun corridor", received their water directly from it. Their close proximity to the intake, and position along the transmission main, meant that they were able to easily secure a supply and draw down on the available water, without appreciation for its value (Figure 10.11) and the huge demand that was building in and around the region's growing capital city. Water losses and revenue collections were becoming an increasing problem in these areas, which resulted in significant water shortage issues to address and a lack of funds to enable improvement works without external funding support.

The expanding population of the Tamale metropolis resulted in a rationing program whereby the city was split into three zones, each nominally receiving water for two to three days a week. In practice, the massive water shortages resulted in a further breakdown of the rationing zones, with any given part of the city only receiving water for a few hours on a single day each week.

A project scope was prepared, with finance (part grant, part loan) arranged through the Development Related Export Transaction Programme (ORET) of the Dutch government and the Dutch Development Bank (FMO), to expand access to water and improve the service level of the existing network to provide continuous supply. The scope of the project was to extend the capacity of the raw water intake and treatment plant to 44 MLD, construct a new 700 mm transmission line to a newly constructed reservoir located on high ground at Kukuo Yepalsi, just to the north of Tamale, from where water would gravitate to the city, and the construction of 96 kilometres of distribution pipe.

Figure 10.11 Water wastage along the Dalun corridor.

The project included a Technical Assistance Program to build capacity of the local utility to help ensure the sustainability and optimisation of the capital investment. This included the establishment of a water distribution management centre in the local utility headquarters (Figure 10.12), the installation of meters and creation of District Metered Areas (DMAs) throughout the network, creation of a Geographic Information System (GIS), network model, per capita consumption studies, pressure management investigations, leakage and Non-Revenue Water (NRW) control and the creation of a competent local utility operations team.

Figure 10.12 Water Distribution Management Centre at Regional GWCL Headquarters.

To address the specific issues in the Dalun corridor a tri-sector partnership was established with the contractor, the utility and a coalition of local Non-Governmental Organisations (NGOs) or partner organisations who were led by the Netherlands Development Organisation (SNV). The tri-sector partnership was tasked with engaging with communities along the corridor to educate them on water issues, get them to understand the value of water, encourage them to maintain the water network, reporting any deficiencies to the utility and above all, encourage them to pay their water bills.

10.3.2 Challenges of intermittent water supply

Implementing the water-rationing regime involved a never ending series of valve operations, over 100 per week. This placed an inordinate burden on the utility maintenance team. The severe restrictions meant that the official water rationing times were little more than a guide. Commendably, the utility did manage to mostly stick to the scheduled days served. However, it was impossible to predict the supply time with any confidence. This meant that customers had to be on standby on their supply days to ensure that they received sufficient water during their limited supply time to meet their needs for the week (Figure 10.13).

Figure 10.13 Queuing and storing water under the intermittent supply regime.

The alternatives available included receiving water from local vendors, which was expensive, queuing at a local standpipe, which could take a long time, or taking water from an unimproved source, which posed potential health risks (Figure 10.14).

Figure 10.14 Obtaining water from unimproved sources.

The intermittent regime led the utility to throttle multiple valves to prolong the period of supply to each area by restricting the flow. Unfortunately, this constant throttling placed the valves under considerable strain causing them to deteriorate.

The limited and uncertain hours of supply, combined with the low pressures, made active leak detection in the city practically impossible under the rationing regime. Trying to assemble a utility leak detection team to coincide with the hours of supply was akin to chasing shadows. Whenever it was achieved, the throttled valves, low pressures and inadequate zoning made conventional leak detection ineffective.

The network along the high-pressure transmission line to the city posed its own challenges. Communities were well aware that this was the only pipe from which a reliable water supply could be obtained. As such there was intense demand to connect to it. This resulted in air valves and washouts located along the line being improvised as network distribution offtakes (Figure 10.15). Two communities almost came to civil conflict when one community attempted to lay a pipe across another's land to connect to the transmission line.

Figure 10.15 Air valve improvised as an offtake along the Dalun corridor.

As mentioned above, revenue collection was very poor in the Dalun corridor. When the utility threatened a corridor village with disconnection, the local community responded by threatening to vandalise the transmission pipe thereby cutting of the entire water supply to Tamale, forcing the utility to back down. Another problem was the uncontrolled extension of the connections from the transmission main, which in many cases expanded into a spider's web of connections serving ever more distant communities, until pipe head losses and leakage meant the water supply could reach no further. The quality of pipe construction was variable for these offtakes. Few as built drawings or any records of the pipe location, size or materials existed for reference and support with troubleshooting and general maintenance. This further exacerbated the water restrictions to the city.

10.3.3 Intermittent water supply and water losses

The lack of bulk metering, both along the Dalun corridor and in the city of Tamale, meant that it was difficult to directly measure the volume of water loss in different parts of the network. In addition, as the pipes throughout the city were often experiencing dry or unpressurised reduced flows, water loss volumes were limited and it was difficult to accurately assess the location and severity of any leaks. Upon establishing a continuous supply, increasing the flow of potable water throughout the network, the leaks could then be identified and addressed.

On the other hand, in the Dalun corridor, one only had to observe the vegetation along the rural pipelines to detect multiple leakages (Figure 10.16). On one particular two-kilometre offtake literally every joint along the polyvinylchloride (PVC) pipe was leaking. Such losses only served to exacerbate the chronic shortage of water in Tamale.

Figure 10.16 Leakage along the Dalun corridor.

10.3.4 Transitioning from intermittent water supply to 24 × 7

To expand customer access to water and achieve 24 × 7 supply across the entire catchment, the Ghana Water Company lead a multi-pronged approach to address issues throughout the network. Firstly, water production capacity was increased from 19 to 44 MLD. As outlined above, without a determined effort to tackle water losses, such supply augmentation would be totally ineffective and represent poor value for money, so an NRW reduction programme was integral to the initial $50 million water supply contract.

To deal with the particular situation in the Dalun corridor, three specific actions were taken. Firstly, the new 700 mm transmission main was constructed, to which strictly no connections or offtakes were permitted, to allow water to flow directly to the newly constructed Kukuo Yepalsi reservoir (Figure 10.17) and relieve water shortages in the city. Meanwhile, villages along the corridor continued to be fed by the old 450 mm transmission main, which was also extended in order to transmit water to the new reservoir. As such the water supply to the corridor could be better controlled, whilst not jeopardizing supply to the city of Tamale.

Secondly, every offtake and sub-offtake, from the 450 mm transmission main was metered, creating approximately 40 simple DMAs (Figure 10.18). These DMAs had a single inlet and no boundary valves. The creation of these DMAs established clear empirical evidence for the chronic water losses along the corridor and identified those DMAs in immediate need of leak repair or pipe replacement.

Thirdly, the tri-sector partnership, combining contractor expertise, utility experience and partner organisation local knowledge, established water and sanitation committees in selected communities along the corridor with the support of village elders. These combined efforts were able to drastically reduce NRW along the Dalun corridor during the water supply contract, and combined with the increase in production, resulting in substantially more water being made available to serve the people of Tamale.

Figure 10.17 Newly constructed reservoir at Kukuo Yepalsi.

Figure 10.18 Downloading data from DMA meter along the Dalun corridor.

In the city of Tamale, the network was divided into approximately 30 proposed DMAs prior to the augmentation of supply. Meters, boundary valves and pressure monitoring points were installed (Figure 10.19). In some intermittent supply cases the "wet/dry test" was used as an alternative to a proper DMA isolation pressure zero test. In Tamale however, the chronic lack of water, combined with lack of confidence in the network connectivity, prevented the DMAs from being isolated under the rationing regime. The establishment of 40 DMAs in the Dalun Corridor did however enable the logic and management processes of DMA establishment and analysis to be ingrained in the utility prior to the full establishment of continuous supply. Monthly reports were prepared which ranked and prioritised these 40 rural DMAs in terms of NRW expressed as litres per kilometres per day. Targeted leak detection and repair with utility staff then took place in prioritised DMAs.

Figure 10.19 Water monitoring and leak detection in the Dalun corridor.

Following the increase in water production, and the construction of the new infrastructure, continuous 24×7 supply was established throughout the water supply catchment. While NRW initially increased within the city as expected with increased pressure and flows being directed through the existing network, the improved management of water in the Dalun corridor meant that this was not the case in the catchment as a whole. The DMA isolation process began in earnest in the city following the establishment of continuous supply to address the established leaks. However, this still proved quite challenging due to outstanding uncertainty about the network connectivity. Although the majority of pipework in the city was captured on as built drawings and thus transferred to the newly created GIS system, there were many old pipes for which such records did not exist (Figure 10.20).

Figure 10.20 Surveying the water network in Tamale town.

Another problem was that construction of distribution pipe work was limited to just 96 kilometres under the project scope. This brought the total network length to approximately 580 kilometres and 70% customer connectivity. However, there was a clear need, and now sufficient water resources, to serve large sections of Tamale and its environs, which were still not connected to the water supply. Many old existing pipes were in need of replacement both in the city and in the Dalun corridor where certain DMAs continued to have very high losses. The wear and tear that resulted from the continued throttling of valves meant that many were in need of replacement. Unfortunately, there was no provision in the water supply contract scope for such necessary pipe or valve replacement. The utility also had limited resources and materials to deal with leak repairs. On the up side, what the utility did have was much better knowledge of their network together with a desire to make ongoing improvements and they had the skills to manage repair works if additional resources and materials could be secured post-contract.

Coincidentally, during the project the control of utility operations was handed over to a private management contractor, Aqua Vitens Rand Limited (AVRL). Ghana Water Company staff was thus seconded to AVRL throughout the project. The management contractor had a number of performance targets, among them ambitious NRW reduction targets. Following the establishment of continuous water supply the management contractor took the decision to reduce production and thus reintroduce intermittent supply, albeit not at the drastic levels that existed previously, in an attempt to reduce NRW levels and so hit their contractual targets. The management contractor was in a difficult position in that their contract did not include any funds for capital expenditure. This remained the responsibility of asset owner, the Ghana Water Company. As such there was no budget among any interested party to replace defective pipes that would have further reduced NRW and improved the level of service to customers.

Eventually a "solution" presented itself when another external funding agency, the Millennium Challenge Corporation, which stepped in with grant aid of $5 million to scope a further Tamale Water Extension Project executed by the Millennium Development Authority (MiDA). This project extended the network length to approximately 650 kilometres and 79% customer connectivity. This was followed by a World Bank funded Subsequent Year Investment Project (SYIP) which extended the network to 794 kilometres and 94% connectivity.

The current situation in Tamale is that 6% of the population are still not connected to the network. This percentage of the population continues to get water from unimproved sources. For the 94% connected, despite the undoubted improvements, intermittent supply still exists. The old part of the water treatment plant at Dalun is in need of rehabilitation. Efficiency problems have been encountered with the raw water pumps at Nawuni. Power supply interruptions are a significant problem. From September 2015 to April 2016, there was an average of five power supply interruptions per day. There are no electricity generators at the treatment plant. These issues result in the treatment plant operating at below capacity.

There is still a significant leakage problem around the network, especially on polyvinylchloride (PVC) pipes. Some pipes are in need of replacement. Many of the DMA meters need servicing and spare parts. The utility is still challenged to put sufficient resources in place to rectify these problems.

10.3.5 Lessons learnt

The first lesson is that a community will do whatever it needs to secure access to water. Throughout the network, the pipeline breaches, which resulted in an increased number of off-takes along the length of the existing 450 mm pipeline, heighted the acute water scarcity issues faced by those in the city as flows were diminished to all-time lows.

External financial assistance can go a long way in supporting utilities with much needed improvement works, which they wouldn't have been able to afford otherwise. However, as clear in this case, the capital investment provided only goes so far. Subsequent works, including the provision of operation and maintenance resources, is important to sustain and build on the benefits.

With the improved production capacity there should be sufficient water to ensure continuous water supply to all. However, a combination of the following issues continues to prevent this from happening:

- Lack of resources for DMA preventive maintenance
- Lack of resources for leak detection and repair
- Some poor quality network installations
- Power supply interruptions
- Treatment plant in need of rehabilitation
- Raw water pump inefficiencies

Technically it is not difficult to rectify the above issues. The challenge is securing additional funding and resources to carry out the works required and implement a rigorous operation and maintenance effort.

For a project to achieve and sustain continuous supply to the entire population of Tamale, it would need to include the following components:

- Assess pipes in need of rehabilitation
- Replace pipes and extend network to achieve full coverage
- Ensure full DMA coverage of expanded network identifying all cross connections
- Preventive maintenance program to ensure DMAs remain operable
- Properly resourced active leak detection and repair program
- Resources to allow the utility to continuously improve network operations

The above actions should have higher priority than production capacity issues.

10.3.6 Project characteristics

Project: Tamale Water Supply Project
Client: Ghana Water Company Limited
Location: Tamale, Ghana
Executed by: Biwater Contracting BV
Project Value: $50 million
Project Funding: FMO and ORET, Netherlands
Project Duration: 2006–2009
Population Served: 400,000
Production Capacity: 44 MLD
Number of connections: 18,000
Length of Network: 585 km
% Connectivity after project: 70%

This project was almost immediately followed by another, which further extended the network with the addition of 65 km of pipeline and 5,000 new connections.

Project: Tamale Water Supply Extension Project
Client: Ghana Water Company Limited
Location: Tamale, Ghana
Executed by: Millennium Development Authority
Project Value: $5 million

Project Funding: Millennium Challenge Corporation
Project Duration: 2010–2012
Population Served: 400,000
Production Capacity: 44 MLD
Number of connections: 23,000
Length of Network: 650 km
% Connectivity after project: 79%

In addition, a subsequent project was implemented by the World Bank funded 'Subsequent Year Investment Project' (SYIP), which brought the network length to an overall 794 km, connecting approximately 94% of the population and providing a total of 28,556 customer connections (as of December 2015).

10.4 ITALY: ACTION PLAN TO FACE WATER CRISIS – MOVING FROM IWS TO 24 × 7

Alessandro Bettin and Piergiorgio Malano

Keywords: Intermittent water supply; Water loss; District metering; Pressure management; Air control; Pressure transient

10.4.1 Introduction

Due to increasing water scarcity, increasing population and wide urbanisation, Intermittent Water Supply (IWS) is affecting a continuously increasing number of water networks worldwide even in countries that IWS was not at all common. The combined effect of climate change and increasing water demand accelerate the depletion of water recourses. Moreover, due to the changing in precipitation regimes also caused by climate change, the periods of droughts are more and more frequent even in areas characterised by temperate climate. This "new" environment pushes water managers to face unprecedented emergency situations. Most of the time, they face this situation through the interruption of the water distribution for some hours per day in order to allow the reservoirs to be safely filled. IWS is seen as a temporary situation before the implementation of structural interventions that takes longer time and generally bigger investments. Unfortunately, once implemented IWS, to bring back the network to 24 × 7 is really challenging. IWS, in fact, due to continuous and uncontrolled valve operation, determines the increasing of burst frequency and the rapid deterioration of the pipe integrity. Network charging and discharging generates pressure transient and big volume of air along the flow. Both these phenomena are extremely negative for asset duration and network efficiency.

The paradox of water leakage, that in many cases exceeds 50% of the inlet volume, is more and more evident because faced with the increasing demand and water scarcity the traditional approach is directed to exploit new water provisions, new wells, new treatment plants without focusing on network efficiency and leakage reduction.

Valle Umbra Servizi SpA (VUS) is the water utility that manages water and wastewater services in several administrations in the area of Foligno, Spoleto and Valnerina in the Perugia province, in the center of Italy. VUS manages 2.910 km of distribution network serving 152,300 inhabitants. Norcia Municipality has a resident population of around 5,000 inhabitants, with a high number of tourists. The water is mainly supplied by the Capregna source located near the town. Due to modification of precipitation regimes and to the frequent occurrence of drought periods, the availability of fresh water from this source has significantly decreased, during the last years, mainly in the summer season. From an average annual flow of 45 l/s, the flow measured in 2012 was only 19 l/s. In order to meet the increasing demand above all during summer, VUS management decided to operate intermittently turning off the water overnight.

VUS, in the context of the severe water crisis experienced in the early summer of 2012, identified the need to commission an urgent assignment for the hydraulic analysis of the water distribution system of Norcia (Perugia), in order to plan interventions aimed to guarantee an adequate level of service able to satisfy the significant increase of water consumption due to peak tourism. Hitachi Aulos has been awarded the assignment to carry out the study and to deliver a plan of interventions.

10.4.2 The current situation of the water distribution network – initial audit

Following acquisition of the old non-updated network maps and meetings with local water operators, a clear picture of the situation has been established. In order to check the reliability of the existing map and to update the pipe database, a comprehensive survey of the network was carried out. As a result, a detailed hydraulic scheme of the network showing main connections, boundary valves, reservoirs, wells, sources and distribution area was produced. Once the hydraulic scheme was completed and the network operation understood, a water audit was performed. The inlet flow was measured and the water consumption estimated using the data from the billing department of VUS.

As can be seen from the hydraulic scheme (Figure 10.21), from the Capregna source an average flow of 21 l/s is delivered to a distribution tank. From the intermediate tank an average flow of 10 l/s is delivered to the Norcia Reservoir and 7 l/s to Monte Oro Reservoir. 4 l/s is distributed directly from the intermediate tank to a separated DMA with 75 users. The main reservoir is also fed by wells (Fontevena) with an average flow of 9 l/s. The total distributed flow is therefore 34 l/s (including 1.5 l/s from a well plus 1.5 l/s from another well dedicated to the industrial area district). The initial water balance highlighted that the NRW was around 22 l/s (64% of the System Input Volume).

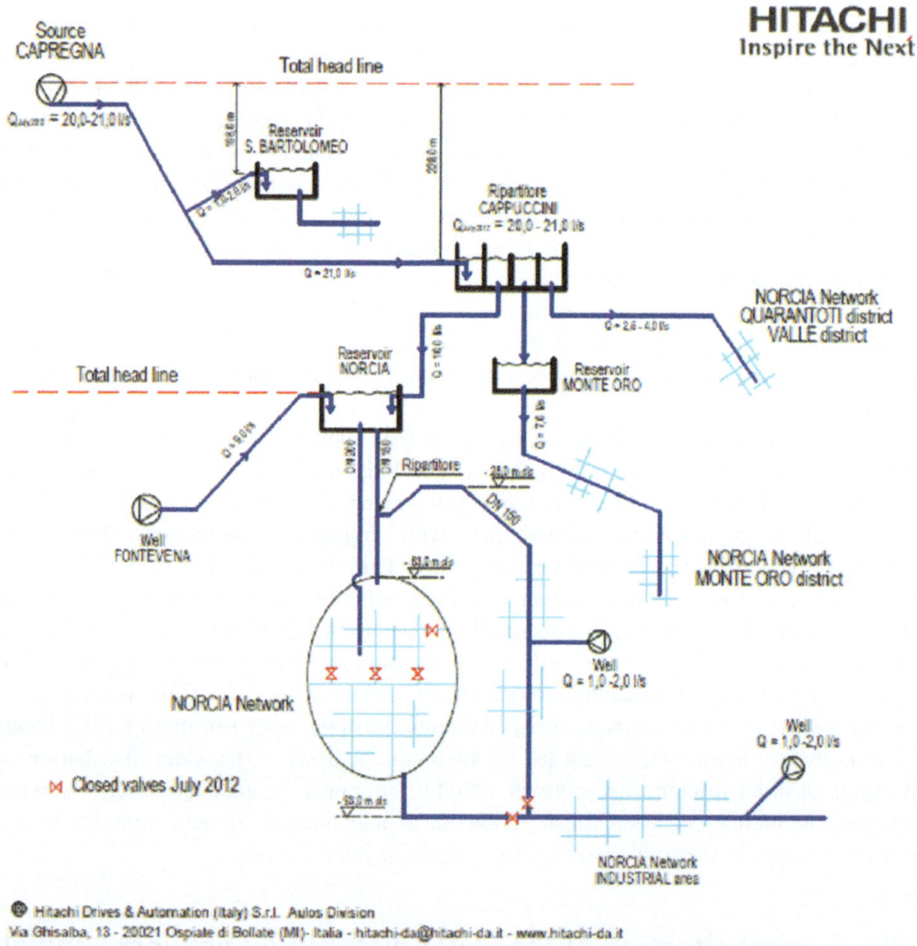

Figure 10.21 The hydraulic scheme of norcia water network. from source and wells to reservoirs and distribution. highlighted in red the boundary valves that were closed to isolate 3 DMAs.

In order to allow the Norcia Reservoir to be filled in the early morning before the peak consumption, the distribution pipes (DN150 and DN200) from the reservoir were usually closed overnight from 00:00 to 06:00 am (Figure 10.22). This kind of operation was generating continuous complaints from customers dealing with "dirty water" due to sediment removal and transport. In some cases the sediment was even generating clogging of meters. Many customers were also complaining for high quantity of pressurized air exiting from their taps and generating vibration to their plumbing system mainly during the morning after the opening of the distribution valves downstream of the storage reservoirs.

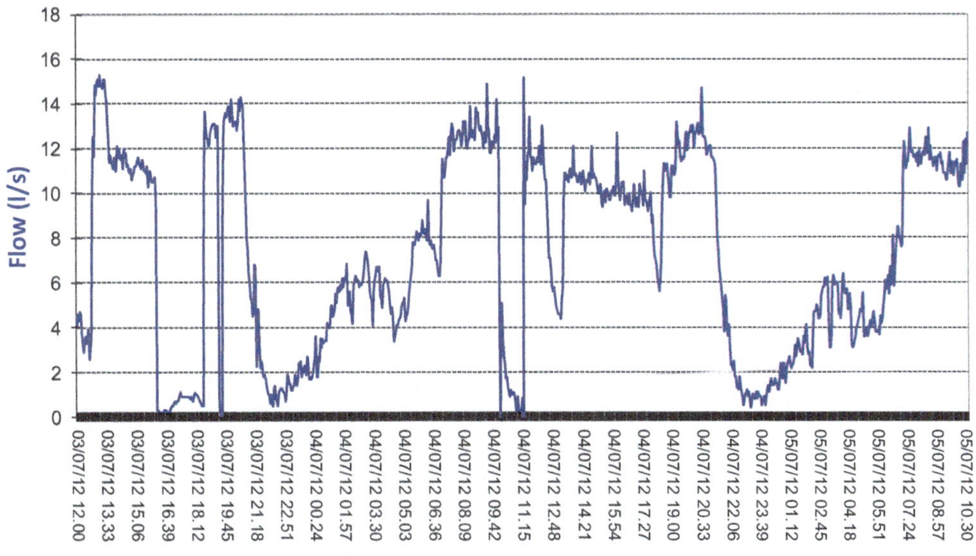

Figure 10.22 Initial flow at the inlet of DMA 2 (DN 200) before intervention. IWS can be seen during night when flow drop to zero.

VUS reported frequent pipe bursts and a high level of water loss that was difficult to locate. The continuous operation of valves to open/close the water supply in a context characterized by uncontrolled pressure is the best environment for the increase of pipe breaks and of the frequency of bursts. Moreover in such a contest the sole repair of leaks is not worthwhile. In fact, after the intervention, leakage generally returns back to the initial level in a short time due to the high leakage rate or rise typical of such cases.

The measurement of leakage level has been obtained by analyzing the Minimum Night Flow after assuring that the valves at the outlet of the reservoirs were open. Physical losses have been calculated as the difference between MNF and legitimate night consumption, real losses were calculated at 19 l/s.

Due to the presence of air in the distribution pipes it was initially difficult to obtain a reliable measurement of the flow at the outlet of the reservoir. The measurement of pressure transient using special data loggers with a recording time step of 5/100 highlighted the presence of pressure transients along the network. Negative pressure has been measured at one main hydraulic node, named "Ripartitore" (Figure 10.23), located downstream the reservoir at –28 m from the piezometric line, in the DN150 feeding the lower area. As can be seen from the picture below, the negative pressure (related to atmospheric pressure) is equal to –7 m. In parallel with negative pressure, due to pressure transient, negative flow was also recorded from the lower node (located at –93 m from the piezometric line) to the reservoir. This situation, probably exacerbated by the uncontrolled opening/closing of the reservoir outlet, was resulting in continuous water hammer highly detrimental to the integrity of the system and eventually responsible for the progressive increase of the number of bursts and pipe collapses.

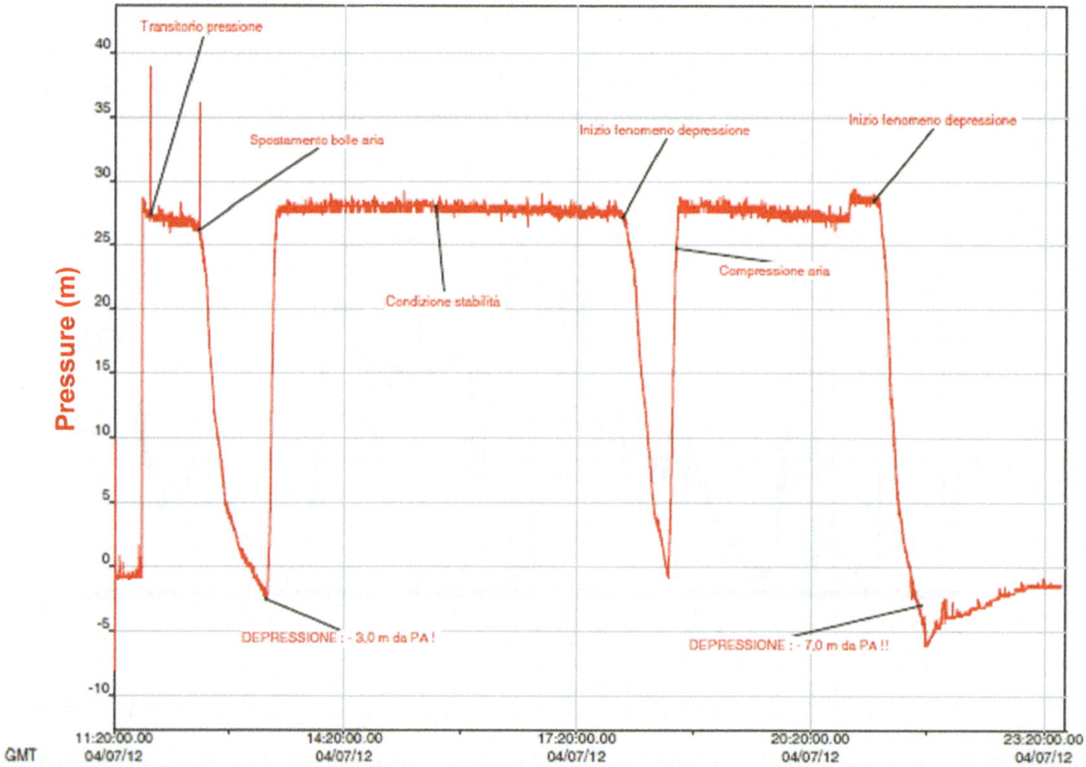

Figure 10.23 Measurement of pressure transient at the node "Ripartitore". The negative pressure highlights the presence of an air pocket subject to compression decompression phenomena. Large quantities of air probably affect even the direction of flow that has been measured negative in the pipe exiting the reservoir, (i.e. flow from downstream to upstream).

In order to limit the detrimental effect due to the air in the pipe network and to reduce at the same time pressure transient an air control valve has been installed at the critical node named ("Ripartitore"). The valve has a threefold function: air venting, evacuation of large volumes of air and allowance of air into the pipe to avoid negative pressure.

10.4.3 Activities carried out

Once the initial audit was carried out, the main effort was directed to the optimization of the network in view of the permanent leakage reduction. To achieve this important result, a direct action aimed at the neutralization of the causes generating pipe burst has been implemented.

The first step was the design of permanent DMA in order to allow optimal pressure management and leakage control. The design was initially carried out on the map and then verified in the field through the direct checking of boundary valves water tightness.

Three DMAs were created closing the valves shown in red in Figure 10.21.

The second step was characterized by the detailed study of pressure control. Field measurements showed high and unnecessary values of pressure particularly overnight. It was therefore decided to installed 2 Pressure Reducing Valves (PRVs) at the inlet of each of the two DMA fed by the Norcia Reservoir (one on the DN80 (Figure 10.24) and the other on the DN200 (Figure 10.25) as highlighted in Figure 10.21). The PRVs selected are equipped with double pilot in order to allow two different settings, day and night with the night pressure set-point lower than the daytime.

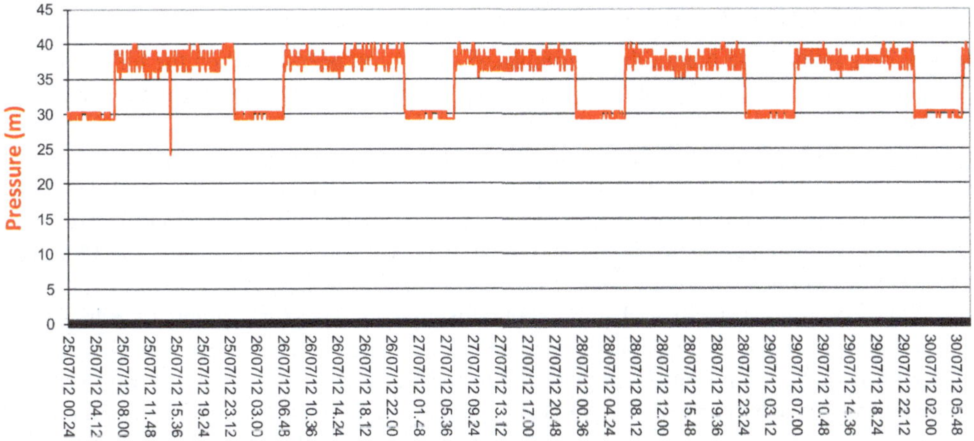

Figure 10.24 Pressure available downstream the PRV installed on the DN80 at the inlet of DMA1 (ground elevation equal to −53 m from reservoir maximum water level).

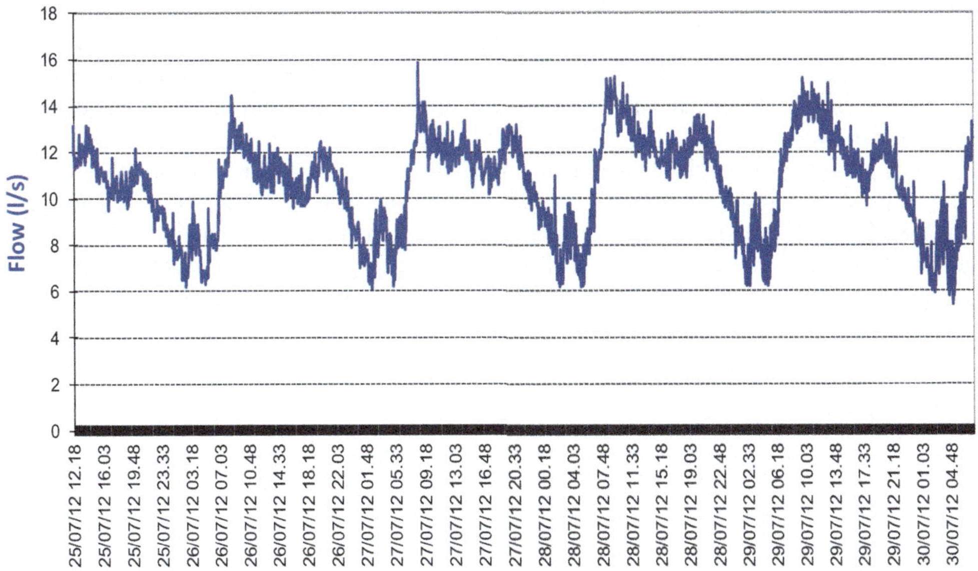

Figure 10.25 Flow at the inlet of DMA 2 (DN 200) after installation of PRV. MNF has dropped to 6 l/s from 10 l/s. Before installation of PRV maximum pressure.

Once the PRVs were installed, leakage detection was carried out starting with the execution of step test to identify most critical areas. Leaks were then been pinpointed using acoustic equipment such as ground microphones and correlators. A total of 20 leaks were located and fixed. In some challenging cases, leaks were pinpointed using tracer gas technology with blend of Hydrogen and Nitrogen.

After installation of PRV, to further optimize network performance, a suitable number of air valves were installed in critical points of the network.

10.4.4 Transitioning from IWS to 24 × 7

The methodology followed to obtain the continuity of supply has been based on the following main key points: measurement of initial hydraulic conditions, analysis, set-up of intervention, maintenance of achieved results.

The first step of the intervention has been directed to eliminate air pockets that were affecting the network performance. The installation of a suitable air valve in the right position downstream the reservoir near the "Partitore" has solved the big part of the problem. Secondly a good margin of improvement was achieved from pressure management. To this end separate pressure areas were created and PRVs installed. Once the network has been put under suitable operation conditions, leak detection has been carried out focusing on the most critical sub districts. Following this process, 24 × 7 was gradually achieved.

At the end of operations and installation, the guideline for the telemetry system of pressure and flow integrated with reservoirs level has been provided to the water company. The operation procedures to maintain over time the achieved results have also been delivered. Thanks to the intervention carried out and the willingness of the water managers to follow these simple but effective rules, today the network is still delivering water 24 × 7 even during the summer peaks and during drought periods.

10.4.5 Lesson learnt

The main effort in any NRW study should be focused on the permanent elimination of the causes that contribute to generate new leaks and pipe breaks. A network affected by intermittent supply is generally stressed by the repetitive and uncontrolled opening and closing of the section valves by the operator. Generally this action is done manually and too fast resulting in pressure transients and air pockets travelling in the network.

The best approach is to establish the right operation conditions in the network through pressure management, pressure transient elimination and air control and then move on to reduce leakage further using traditional acoustic equipment. The benefit from leaks repair can be maintained only if the causes that generate leaks are removed. Otherwise, leaks continue to grow at a fast pace cancelling rapidly achieved results.

10.5 SOUTH AFRICA: THE DANGERS OF INTERMITTENT SUPPLY AS A MEASURE TO SAVE WATER

Ronnie S Mckenzie

10.5.1 Abstract

There are many countries around the world where intermittent supply is a way of life and residents often have to deal with a water supply which is only pressurised an hour per day or even an hour per week. Such intermittent water supplies cause huge problems to the long term viability of the water reticulation systems and pose significant health risks.

In South Africa, the issue of reducing water losses from municipal water supply systems is becoming a serious problem throughout the country, particularly in view of the current (2016) drought situation which is aggravating an already difficult balance between supply and demand. Municipalities throughout the country are being asked to cut their water use, many of whom have neither the expertise nor the funds to implement the appropriate interventions with the result that little progress is being made.

It has become clear that some water suppliers are resorting to the introduction of intermittent supply as a measure to reduce losses and normal water consumption. In some instances, there is clearly no alternative due to the fact that the supply reservoirs are at or near empty. In other cases, the practice of intermittent supply is being used as a quick and simple measure to reduce water losses.

The case study will highlight the implications of intermittent supply and the fact that it can be introduced in a matter of hours, while the damage caused to the reticulation system may take years to resolve. The case study will show that although some savings can initially be achieved through the introduction of intermittent water supply, in the long-term, such measures will often result in higher water use.

10.5.2 Introduction

Potable drinking water is becoming one of the most important issues in the 21st century. Growing world population, global warming, improved living standards in many areas and land use changes are among the many factors which exacerbate the impacts of the normal flood and drought events. Climate change is often highlighted as the key factor behind all droughts and floods which often appear more severe than any previously recorded events. While the cause of the more extreme events is often up for some debate, the fact remains that both floods and droughts are becoming more severe and causing huge stress on many regions throughout the world. The situation is unlikely to improve in the foreseeable future and may well deteriorate with the result that water supply managers and other government officials must prepare for more extreme events in future and ensure that the available resources are being used efficiently.

When considering potable water supply, South Africa has many different problem issues to deal with. By global standards, the country has relatively efficient reticulation systems in most of the metros and large municipalities where tap water is generally safe to drink and water is generally supplied at normal pressures 24 hours a day. While this is often taken for granted in many parts of South Africa, it is unusual in most parts of the LAMI world. In many of the rural areas in South Africa, safe drinking water cannot be taken for granted and such areas regularly experience severe supply problems during times of drought. Such local water supply schemes are extremely vulnerable to drought events due to the fact that they often rely on run-of-river flow or groundwater for their water supply. Such small local schemes, lack the assurance of supply provided by the large integrated water supply systems used to supply the urban areas where reservoirs are often linked to each other across provincial boundaries to provide a reliable supply that can withstand droughts of 5 or even 10 years in duration. In effect, rural water supply schemes will tend to experience regular but shorter drought events while the large inter-basin schemes will experience less frequent but potentially much longer drought events.

The above points are well illustrated by some facts and figures which were recently presented by the Department of Water and Sanitation (DWS, 2016) to highlight the water situation in the Vaal River System which supplies the main industrial heartland of South Africa including the whole of Gauteng, many mines, power stations, large industries etc. Figure 10.26 provides a graph showing the percentage storage in the major reservoirs which supply the Vaal River System (based on information from DWS, 2016). The actual storages in each of the main reservoirs in the system as on 1 May 2016 are provided in Table 10.8.

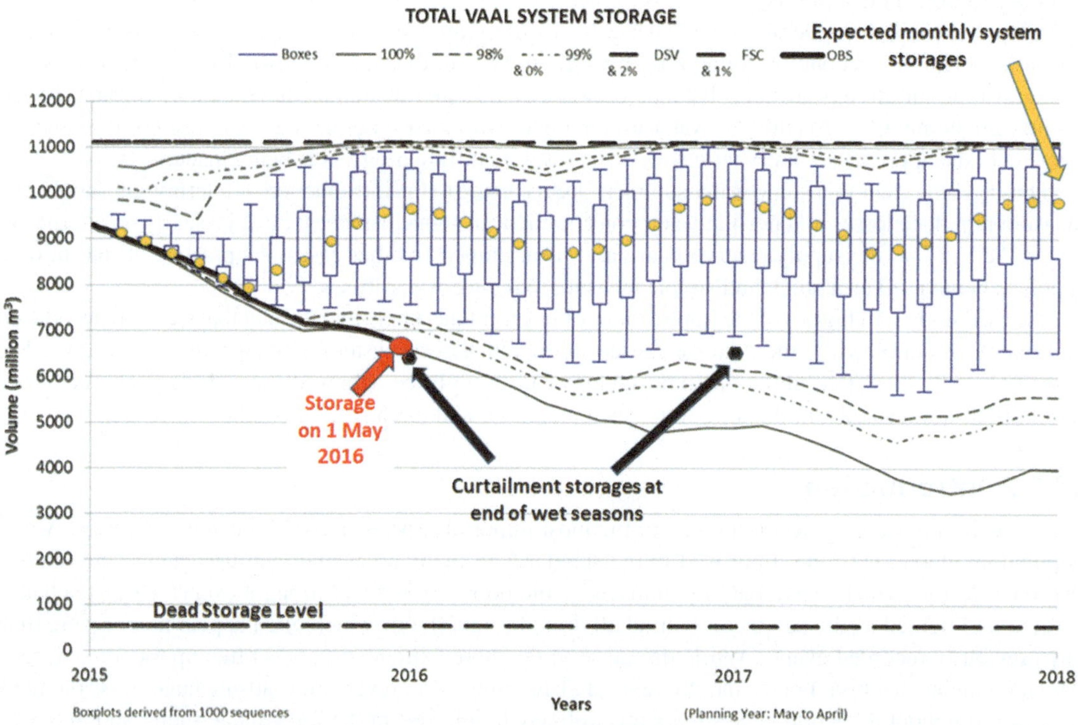

Figure 10.26 Storage status of vaal river system at end of 2015/16 wet season (DWS, 2016).

As can be seen from the figures, the major reservoirs in the Vaal River System at the end of the 2015/16 wet season were around 60% full. In most countries around the world, this would not be considered to be a major concern, however, in the South African context, it is considered to be a significant risk and is very close to the level at which restrictions are considered. In order to operate and manage its water resource systems, the Department of Water and Sanitation utilises, what is widely recognised as one of the most sophisticated water resource models in the world. Rather than waiting for the situation to deteriorate to the extent that has recently been seen in many other countries around the world, the Department of Water and Sanitation, tries to introduce drought restriction measures as early into a drought event as can be identified. The policy of introducing restrictions early in the drought, helps to avoid the introduction of the most severe restrictions which can have a serious and detrimental effect on the economy. In many cases, less severe restrictions applied early in a drought will be sufficient to avoid severe restrictions as has been demonstrated in the Vaal River System over the past 35 years during which only light restrictions have ever been applied.

Table 10.8 Reservoir storages for vaal river system.

Dam Name	Full Supply Volume (mil m^3)	Volume May 2016 (mil m^3)	% Full
Sterkfontein	2,617	2,319	88.6
Vaal	2,610	1,134	43.4
Katse	1,950	1,399	71.8
Bloemhof	1,240	231	18.6
Mohale	947	333	35.2
Heyshope	447	404	90.3
Woodstock	373	328	87.8
Grootdraai	350	313	89.2
Zaaihoek	185	104	56.4
Morgenstond	101	55	54.6
Nooltgedacht	78	53	67.0
Vygeboom	84	54	64.1
Westoe	61	34	56.5
Jericho	60	48	79.6
Totals	11,103	6,808	61.3

10.5.3 Intermittent water supply

Intermittent water supply is the practice of cutting off the water supply into an area at certain times of the day. Typically, it is used to reduce water consumption into an area by cutting off the supply during the night-time period when consumers are asleep and not using water in which case the water supply is cut from approximately 8 pm to 5 am which will normally be achieved by manual closing of valves or through the use of electronic devices which can be programmed to open or close the valves at specific times of the day.

Intermittent supply is usually introduced either as an emergency measure, when the supply reservoir is close to, or in some cases it is introduced as a measure to control water use and to reduce wastage. In the first case when there is little water left in the supply reservoir, there may be no alternative to the rationing of water and an intermittent supply cannot be avoided once the supply resource has been depleted. In the second case, where the intermittent supply is introduced as a water saving measure, however, there may well be alternative interventions that can provide savings without some of the problems that tend to accompany such pressurising and depressurising of the system. In South Africa, many municipalities are currently considering intermittent supply as a semi-permanent water saving measure. The remainder of the case study addresses some of the important issues that should be taken into account when introducing such measures.

10.5.4 Health issues

Before looking at the actual water savings that can be achieved through intermittent supply, it is important to recognise the potential health risks associated with such practice.

Figure 10.27 provides a simplified layout of the water supply to a property, which has either water borne sewage or a basic septic tank system. In this example, the sewer network is shown to be leaking with the result that there is some level of sewage contamination in the ground surrounding the water supply pipes. Under normal conditions, the water supply is fully pressurised and therefore any leakage from the water supply network is effectively clean water leaving the pipes and mixing with the surrounding groundwater which may or may not be contaminated. As long as the system remains pressurised there is minimal opportunity for the water supply to the property to be contaminated in any way.

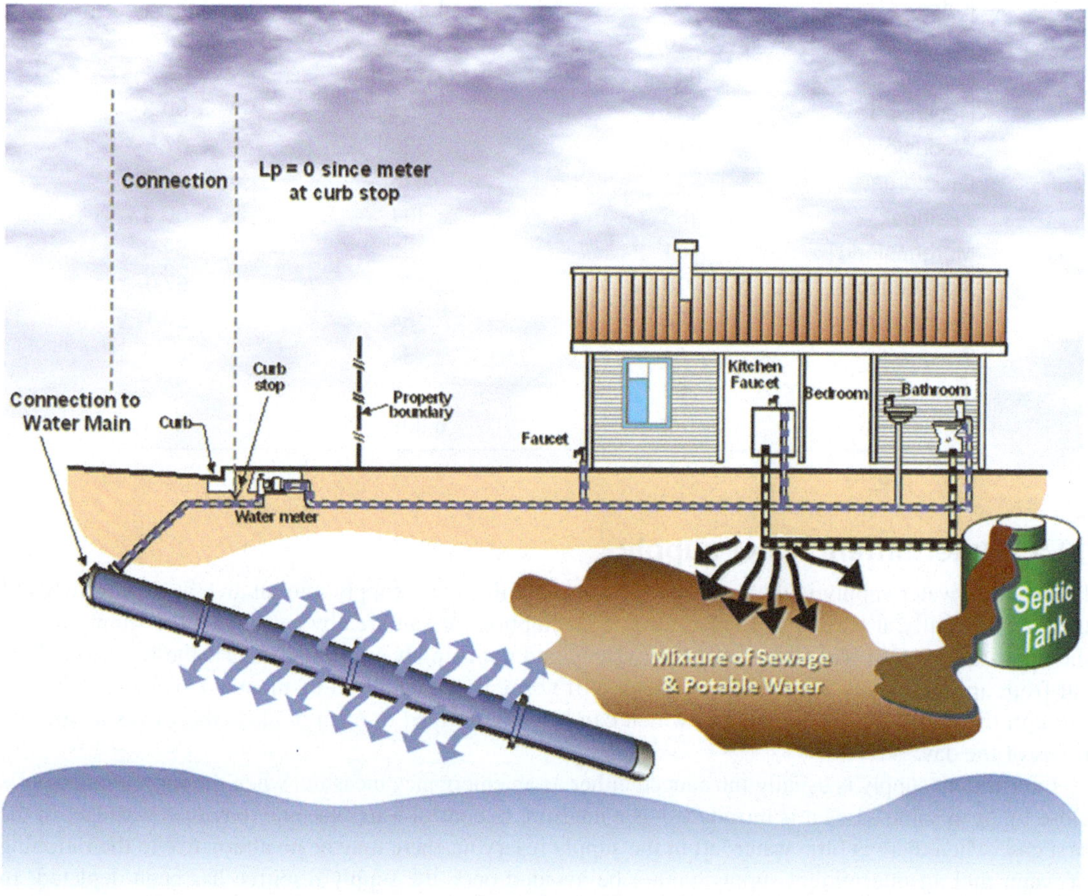

Figure 10.27 Typical water supply to a property when pressurized.

In the event that the water supply system is depressurised as shown in Figure 10.28, there is a possibility for the groundwater surrounding the water supply pipe to enter the supply since any leaks in the pipeline can allow water to move in either direction. If the groundwater surrounding the pipes is contaminated in any way, there is a risk that contaminated water will be supplied to the property when the system is re-pressurised as shown in Figure 10.29.

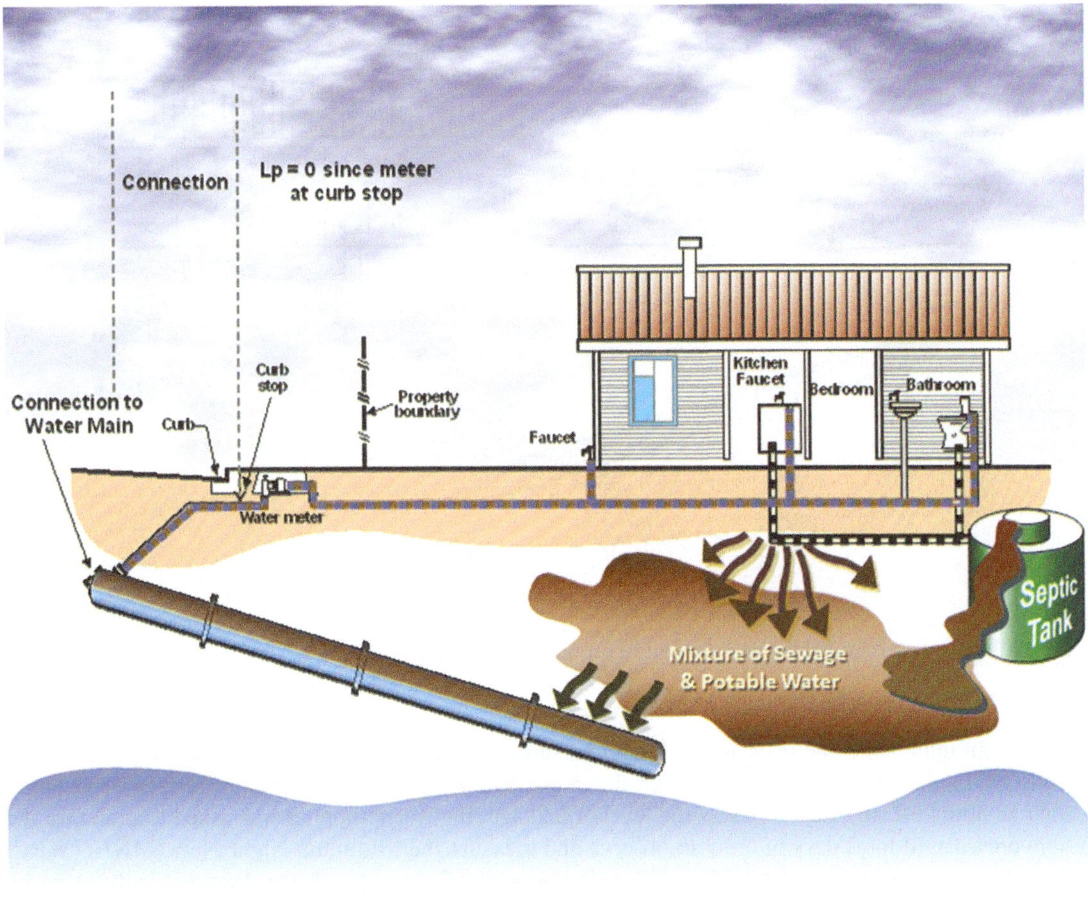

Figure 10.28 Water supply contamination when system is depressurised.

The above figures highlight the potential health risks associated with intermittent supply which can result in certain water borne diseases in cases where there is insufficient chlorine in the potable water to eliminate the contaminants. In such cases, typhoid and cholera can occur.

10.5.5 Reducing consumption through intermittent supply

The main purpose of introducing intermittent supply is to reduce water consumption. Figure 10.30 depicts the flow logging into a residential area in South Africa in which intermittent supply has been introduced by the municipality in an attempt to save water.

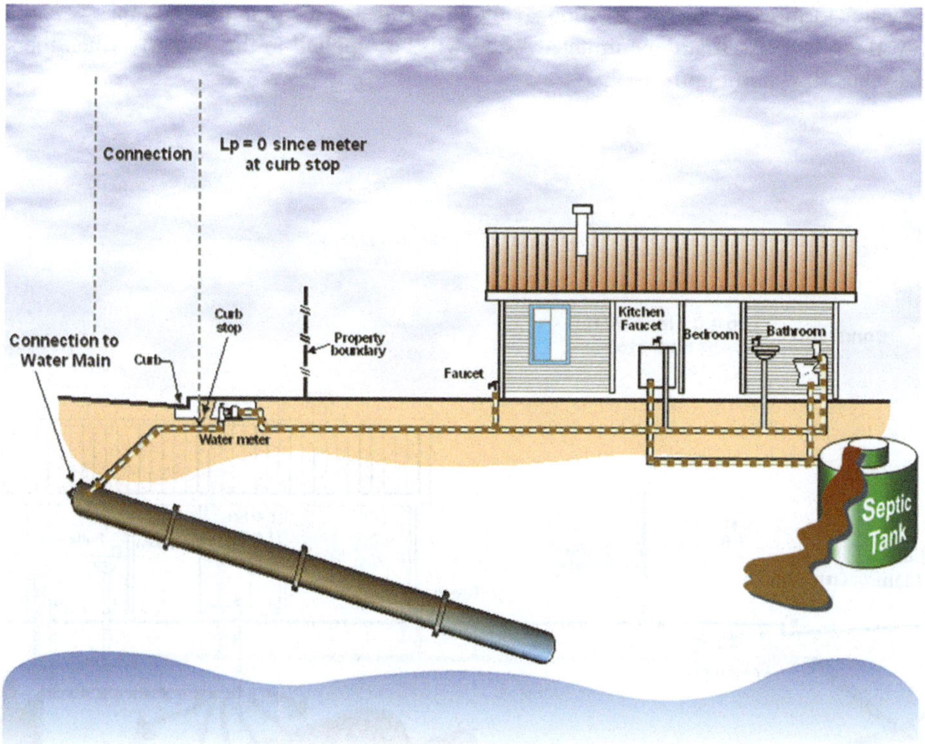

Figure 10.29 Contaminated water supplied to property when re-pressurised.

This is an interesting and informative logging result from an area which was experiencing high levels of household leakage. In order to save water, the municipality adopted a water supply regime of 12 hours on and 12 hours off. In order to assess the level of leakage, the water supplier was asked to maintain the system pressurised for 2 days in order to observe and measure the Minimum Night Flow (MNF) (WRC, 1999) which was estimated to be 165 m^3/hr. Over the same 2 day period, the Average Daily Demand (ADD) was measured to be 200 m^3/hr. In such cases, the ratio of the minimum night flow to the average daily demand (MNF/ADD) is a useful indicator of leakage in the area. In this case the MNF/ADD ratio was found to be 0.83 or over 80%. While this calculation is not always a reliable indicator (since not all of the night flow is leakage), it is generally a useful indicator in areas experiencing high levels of leakage. Typically a system with little leakage will have a ratio of around 0.1 to 0.2 and anything higher will suggest some form of leakage problem. In South African systems, any ratio above 0.2 is considered to be an indication of high leakage and any ratio above 0.5 will usually reflect high internal household leakage. Figures in excess of 0.9 have been observed in some areas.

Another interesting issue that can be seen in Figure 10.30 is the refilling spike that occurs each morning when the system is initially re-pressurised. This spike in flow is caused by the initial volume of water required to fill the empty pipes after a period of zero water pressure. Such sudden and extreme high flow rates can damage the water meter if the peak flow exceeds the upper limit of the meter. In addition to the refilling spike, there is also the problem of air in the pipeline which must escape somewhere and if the air-valves on the pipeline are not functional, the air will create problems either on the reticulation or in the consumer properties. If intermittent supply is allowed to continue, the leakage rates in the reticulation system will increase to such an extent that the system will eventually break down completely.

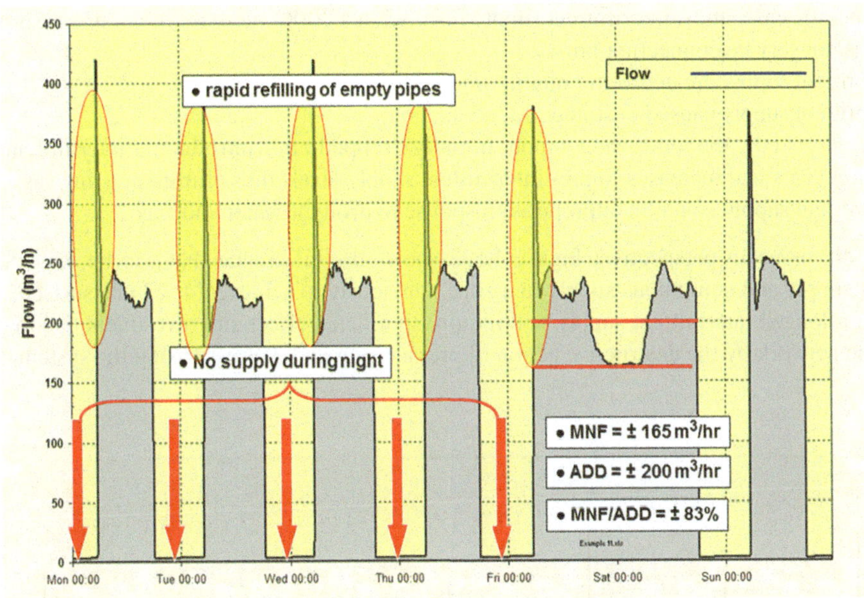

Figure 10.30 Typical flow logging for a zone with intermittent supply.

Figure 10.31 shows an example where a water supplier decides to introduce intermittent supply for 10 hours each night between 8 pm and 6 am. The average daily consumption to the area drops from 17.4 m³/hr to 14.0 m³/hr which represents a reduction of approximately 20% for a 40% reduction in time pressurised. In this example, the leakage is already very high and yet the benefit achieved is relatively modest. In an area with low leakage, the benefit will be significantly smaller and leads to the question of whether or not such a measure is worthwhile.

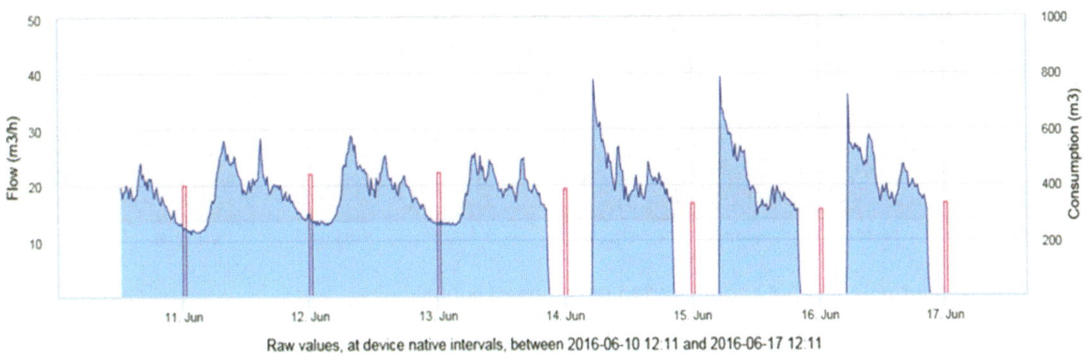

Figure 10.31 Water savings through intermittent supply.

Recent research undertaken internationally (Charalambous, 2016) also raises questions regarding the water saving benefits of introducing intermittent supply. On a large scale project undertaken in Cyprus, Charalambous observed the following:

- Intermittent supply had a detrimental effect on the integrity of the distribution network;
- In a specific case study, intermittent supply resulted in a 200% increase in mains bursts and a 100% increase in service connection bursts;
- Intermittent supply created water quality problems;
- Intermittent supply caused customer dissatisfaction;
- In the case study, the total water supply initially dropped after introducing intermittent supply but within 3 years had increases to pre- intermittent supply levels thus eliminating any savings;
- Intermittent supply is not an appropriate response to drought/water shortage

To place the issue in perspective, Figure 10.32 shows the difference that can be expected between intermittent supply and continuous supply in a zone. Obviously to achieve 24 × 7 pressurised supply after a period of prolonged intermittent supply will require significant effort and investment due to the damage caused to the network by the destructive action of pressurising and depressurising the system.

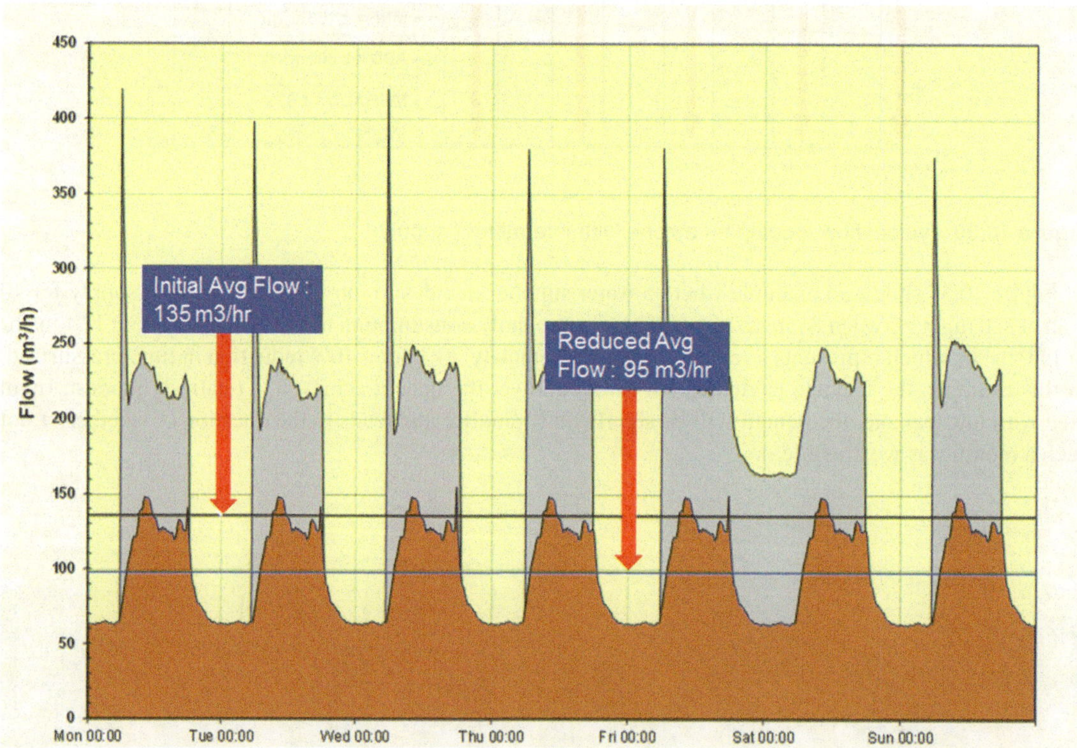

Figure 10.32 Intermittent supply compared to pressurised supply.

The example shown in Figure 10.32 is based on what can be achieved if the system leaks are repaired and the pressure is used as a management tool. As can be seen, the average water use by the zone is 135 m³/hr when operating under intermittent supply and 95 m^3/hr when operating under 24 × 7 pressurised supply. While the figures will naturally change from system to system, the underlying message is clear – intermittent supply will eventually result in greater water use than 24 × 7 pressurised supply assuming that the resource is capable of supporting the demand from the system.

10.5.6 Conclusions

Based on real results from systems around the world the following conclusions can be drawn with regard to the use of intermittent supply when used as an intervention by municipalities to save water:

- Intermittent supply is considered by many municipalities to be a quick and easy intervention for reducing water use in an area;
- Intermittent supply will result in a short-term reduction in water input but this is likely to be offset by increased leakage in the long-term resulting in an overall increase in water use;
- Intermittent supply damages the water reticulation network and increases the number of bursts;
- Intermittent supply can create water quality problems and in some cases increase the incidence of water related diseases such as typhoid and cholera;
- Implementing intermittent supply can be achieved in a matter of hours while returning to a reliable 24 × 7 pressurised supply can take years due to the damage caused to the network;
- Water service providers that have fallen into intermittent supply have major governance and incentive flaws and need in-depth reform; moving to continuous supply requires often very difficult political and institutional choices that many governments prove reluctant to make. A paradigm shift is imperative (Charalambous, 2016).
- Intermittent supply is not an appropriate long-term intervention for reducing water use or water losses.

10.5.7 Recommendations

- Water service providers should implement water saving measures (WRC, 2014) and/or restrictions early on in a drought to avoid having to take emergency measures when water supplies can no longer support the demands;
- Reducing water pressures to below the acceptable norms in order to maintain a 24 × 7 pressurised supply is preferable to introducing an intermittent supply. Liaison with the fire department will be helpful in such cases to address liability and insurance issues;
- Properly sectorised networks with clearly defined and controlled pressure zones have the added benefit that pressure can be used to control water use during severe droughts. Water service providers should therefore be encouraged to set up and operate such zones during normal operating conditions so that the necessary infrastructure is in place when needed.
- Intermittent supply should be the option of last resort and not the option of first choice.

10.6 SRI LANKA: INTERMITTENT WATER SUPPLY AND WATER LOSSES

Joe Dalton

Keywords: Intermittent water supply; District metered areas; Network efficiency; Pipe replacement

10.6.1 Introduction

The town of Negombo in Western Sri Lanka, approximately 30 kilometres north of Colombo, just north of the international airport, is served with water from an abstraction point and treatment plant at Bambukuliya (Figure 10.33) on the Maha Oya River just to the north of Negombo. In addition to the town of Negombo, the treatment plant also serves Colombo International Airport and a Government designated Board of Investment area. Because of the condition of the transmission mains the treatment plant at Bambukuliya was not operating at full capacity.

Figure 10.33 Water Treatment Plant at Bambukuliya.

There was only 59% connectivity to the water network in the town and a supply time average of eight hours per day for those connected. Unconnected households received their water from a combination of sources including private wells and networks, public standpipes and from neighbours with water access.

A feasibility study was prepared to achieve 100% connectivity and continuous 24 × 7 supply to a design horizon of 2025. Using this study, project finance was arranged in the form of a grant from the Dutch Government Development Related Export Transaction Programme (ORET) and a loan from the Dutch Development Bank (FMO). The project scope was to expand the raw water intake and production capacity at Bambukuliya from 24 to 36 mega litres per day (MLD), upgrade the transmission lines (Figure 10.34), construct a new reservoir and install distribution pipe extensions and replacements. The transmission main construction included connection with the Greater Colombo water supply so that the Negombo catchment could be augmented with a further 21 MLD making a total of 57 MLD available to Negombo following project completion.

Figure 10.34 Transmission line from Greater Colombo water supply.

The project included a Technical Assistance Program to build capacity of the local utility (Figure 10.35), optimise operation of the expanded network and sustain the investment benefits. A water distribution management centre was established in the building adjacent to the local utility Operations and Maintenance (O&M) office. From this office a Geographic Information System (GIS) was created and the new network extensions were designed incorporating full District Metered Area (DMA) coverage over the project catchment. A network hydraulic model was built and used to optimise the design of the water distribution system. Extensive training took place of the local utility staff to create a competent network operations team.

Figure 10.35 DMA monitoring in Negombo.

10.6.2 Challenges of intermittent water supply

The intermittent supply situation brought the usual difficulty of carrying out leakage detection when pipes were dry for 16 hours a day. It was not possible to conduct night flow analysis under these conditions. Intermittent supply made it difficult to carry out network zoning. The lack of adequate zoning made prioritisation of areas for detection activities difficult. This was compounded by the utility standard residual network pressure of six metres, which meant that acoustic detection techniques were mostly ineffective. A lack of sufficient network valves made step testing difficult.

The utility was licensed by the Sri Lanka Irrigation Department to abstract water from the Maha Oya River. Some distance upstream from Bambukuliya at Pinnewala was the famous elephant orphanage where elephants came to drink and bath in the river (Figure 10.36).

Figure 10.36 Elephants bathing in the Maha Oya River at Pinnewala.

There has been historic competition in the country between agricultural and domestic users. With the expansion of the intake and treatment plant at Bambukuliya it was not anticipated that any further abstraction volumes would be permitted for the foreseeable future. Establishing and maintaining continuous supply in the town therefore depended on optimisation of the network and minimisation of water losses.

Implementing continuous supply represents a change in how the utility serves its customers. Even though the change is meant to be beneficial, if customers are accustomed to receiving a service a certain way, any change has to be carefully managed. A survey of 54 Negombo households, both connected and unconnected, showed 75% had no idea of plans in place to implement continuous supply, though 94% did express their desire for such a service.

Prior to continuous supply implementation there were a number of public standpipes that provided "free" water for the poor. The bill for this water was meant to be paid for by the Municipal Council. However they had a poor record of paying their bills, reflecting a common trend in LAMI countries of public institutions having high water billing arrears. It was also observed that wealthier people were collecting water from these standpipes. Following continuous supply implementation the utility planned to disconnect these standpipes. However the customer survey showed that users who relied on these standpipes were entirely unaware of such plans. This risked causing public disquiet about a change in service that was meant to be beneficial.

10.6.3 Intermittent water supply and water losses

Under the existing intermittent supply regime, the utility had no active leakage detection teams. They repaired all leakages reported by the public. Such leakages were invariably repaired during the water supply down period. The utility did have an efficient customer meter reading and billing program with working procedures agreed with the unions.

Establishing full DMA coverage over all of Negombo was considered a crucial part of successful continuous supply implementation allowing the prioritisation and targeting of NRW detection efforts. The contractor staff worked with the utility staff in the water distribution management centre to prepare the designs for the network extensions and replacement (Figure 10.37). The breakdown of distribution pipe work in the project catchment is shown in Figure 10.38.

Figure 10.37 Pipe Construction in Negombo.

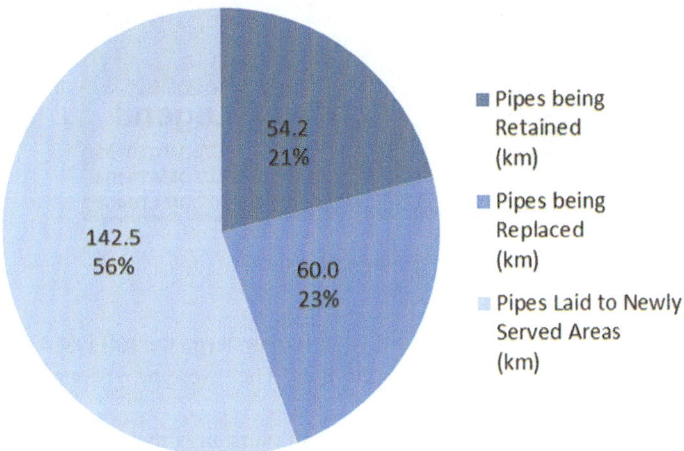

Figure 10.38 Distribution pipe breakdown for Negombo water supply project.

A total of 38 DMA designs were prepared for the entire catchment. For the 21% of the network being retained, the DMA designs incorporated enabling works, including meter, valve and pressure monitoring point installations.

The Utility accepted advice from the contractor to increase the residual design pressure from 6 to 10 metres in order to improve service level to customers and facilitate acoustic based leakage detection techniques.

An extensive customer geo-location process linked the billing to the GIS system and allocated all customers to their proposed DMAs. This facilitated accurate DMA water loss calculations and analysis. This was achieved by working with the utility customer meter reading staff.

Three trial DMAs were identified within the project catchment, the location of which are shown in Figure 10.39.

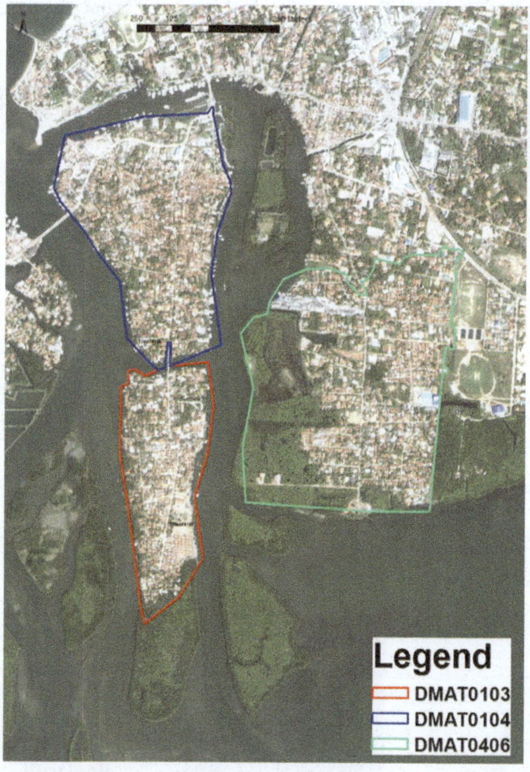

Figure 10.39 Three trial DMAs in Negombo catchment.

These were used as training vehicles for the utility staff to undergo the full DMA establishment process, followed by NRW analysis, detection and repair activities. They were provisionally established under the intermittent supply regime.

As shown in Figure 10.39, two of the trial DMAs were on islands making their isolation very straight forward. The third, 'DMAT0406', was more typical of the type of DMA that would be encountered in Negombo (Figure 10.40).

Figure 10.40 DMA meter installation in Negombo.

10.6.4 Transitioning from intermittent water supply to 24 × 7

Following commissioning of the augmented treatment plant at Bambukuliya, and completion of the transmission and partial completion of the distribution infrastructure, continuous supply was established in 50% and then 75% of the network in June and July 2011 respectively. The construction program stipulated which parts of the network would be supplied first. Initial continuous supply implementation incorporated the three trial DMAs, which were tested for isolation following such implementation (Figure 10.41).

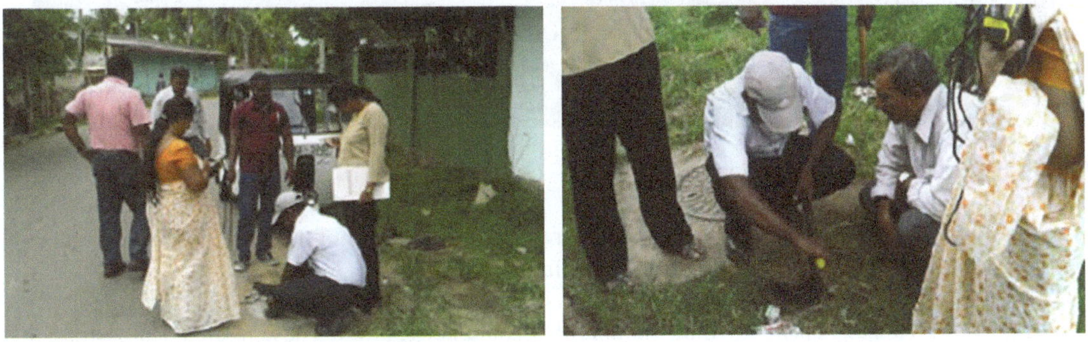

Figure 10.41 Isolation testing of trial DMAs in Negombo.

Due to their natural layout, two of the trial DMAs were easily isolated. However isolating DMAT0406 proved quite challenging due to old cross connections dating back to when the water network was under municipal control several decades before. Finally, after excavating the pipe connectivity junction close to the DMA inlet an additional breach was discovered after which the DMA was successfully isolated (Figure 10.42).

Figure 10.42 Identifying cross connections to trial DMAs.

Following isolation leak detection was carried out in the three trial DMAs by the utility staff (Figure 10.43). Despite the improved supply situation, the pressure profile still limited the effectiveness of acoustic equipment such as leak noise correlators, ground microphones and noise loggers. Step testing proved to be an effective method of leak localisation. This necessitated the installation of additional step valves. Open sounding of all customer meters identified many leakages on the customer connections following continuous supply implementation, which helped to prioritise areas that needed urgent attention.

Figure 10.43 Leak detection in trial DMAs.

Overall throughout the project catchment there was an upsurge of reported leakages following continuous supply implementation as illustrated in Table 10.9.

As shown in Table 10.9, up to June 2011, all leaks that were reported were repaired, but thereafter the number of reported leaks outweighed the number that could be repaired. These figures demonstrate that past a certain point, the utility was under-resourced to deal with the increased number of reported leaks, even without considering active leakage detection. In one particular trial DMA NRW increased from 78 to 256 litres per connection per day upon continuous supply implementation due to the significance of the leakage issues throughout the DMA.

There was also an immediate upsurge, over 100%, in customer billing complaints following continuous supply implementation. The majority of complaints were related to high customer bills and discolouration issues. Discolouration was caused from the increased amount of leakage repairs undertaken (Figure 10.44).

Table 10.9 Leaks reported before and after continuous supply implementation.

Month	% of Network Under Continuous Supply	Leaks Reported	Leaks Repaired
April 2011	0%	145	145
May 2011	0%	268	268
June 2011	50%	468	345
July 2011	75%	428	296

Figure 10.44 Leak repair in trial DMAs.

Complaints relating to high customer bills showed that customers' usage increased. However, they had not considered the costs associated with increased water usage and the resultant bill they would receive.

In the Pitipana isthmus, in the south of the Negombo catchment, low network pressures led to political lobbying for additional cross connections which would have overridden the network designs and jeopardized continuous supply in central Negombo. The solution was to increase the pumping regime from the installed duty, assist standby pumps taking into account the demand curves.

10.6.5 Lessons learnt

The case of Negombo shows a utility dependent on external financial assistance to achieve continuous supply. Such project finance arranged by the Contractor including in this instance a Grant making the finance concessional in nature, offered an affordable solution to the problem of funding such capital-intensive works.

Maintaining continuous supply depends on optimising operation of the network and minimising water losses. Continuous supply creates an additional burden for Operation and Maintenance and initial increase in NRW. The utility, or entity tasked with these NRW works, must rise to this challenge and put resources and budgets in place to deal with this.

The challenge of isolating one of the trial DMAs (DMAT0406) illustrates the information gaps that can exist in water networks in LAMI countries. It clearly illustrates that achieving full DMA operability across the catchment represents a significant challenge to the utility.

The very act of isolating the DMA revealed enhanced knowledge of the network connectivity. With persistence this can be replicated across similar catchments. This will then place the utility in a better position to manage their network and sustain continuous supply.

The case of the Pitipana ithmus highlighted the importance to resist political lobbying when it threatens to override the quality of the service provided.

There is a very strong case for greater customer engagement prior to continuous supply implementation. Customers need to understand that with increased water availability and usage comes higher bills. Through carefully managing their usage and conserving water the increased number of customer complaints could have been reduced.

10.6.6 Project characteristics

Project: Negombo Water Supply Project
Client: National Water Supply and Drainage Board of Sri Lanka
Location: Negombo, Sri Lanka
Executed by: Biwater Contracting BV
Project Value: $53 million
Project Funding: FMO and ORET, Netherlands
Project Duration: 2008–2012
Population Served: 175,000
Production Capacity: 57 MLD
Number of connections: 45,000
Length of Network: 257 kilometres
% Connectivity after project: 100%

10.7 REFERENCES

Charalambous B. (2016). Intermittent Water Supply in Water Distribution Networks. In: Paper presented at Water Loss 2016, the IWA Biennial Conference on Water Losses held at the Lalit Ashok Hotel in Bangalore, India. 31 January to 3 February 2016.

DWS (2016). Development of Annual Operating Rules for the Integrated Vaal River System (IVRS): May 2016.

Lambert A. O., Brown T. G., Takizawa M. and Weimer D. (1999). A review of performance indicators for real losses from water supply systems. *Journal of Water Supply: Research and Technology-AQUA*, **48**(6), 227–237.

WRC (1999). Development of a Standardised Approach to Evaluate Bursts and Background Losses in Water Distribution Systems in South Africa: SANFLOW. Report TT109/99, by R. Mckenzie to the South African Water Research Commission, June 1999. ISBN: 1-86845-490-8.

Index

IWA Publishing's authorised EU representative for General Product Safety Regulations is Diane D'Arras, 15 rue Duret, 75116 Paris, France, e-mail: safety@iwap.co.uk.

Printed and bound by CPI Group (UK) Ltd, Croydon, CR0 4YY
16/06/2026
02140352-0001